Johannes Wirz und Norbert Poeplau

Keeping Bees Simply and Respectfully

IBRA

INTERNATIONAL BEE
RESEARCH ASSOCIATION

NB

Keeping Bees Simply and Respectfully – Apiculture with the Golden Hive
Johannes Wirz and Norbert Poeplau
English translation by Martin Kunz.
English proof-editing by David Heaf.

ISBN: 978-1-913811-03-7
Original version © pala-verlag, 2020.
English version, 2021, jointly published by:
IBRA, 1 Agincourt St, Monmouth, NP25 3DZ, UK. www.ibra.org.uk.
&
Northern Bee Books, Scout Bottom Farm, Mytholmroyd, Hebden Bridge, HX7
5JS, UK. Tel: 01422 882751. www.northernbeebooks.co.uk.

Image credits
Cover photos: Norbert Poeplau
Page 20: based on Ferdinand Gerstung, Der Bien und seine Zucht
Page 21 bottom, 32, 85: Christoph Valentien
Page 25: Achillea, adaptation of a rock painting from the Cueva de la Aranãs
Page 26: based on Ludwig Armbruster, Archiv für Bienenkunde
Page 34, 47, 150 bottom, 165 bottom: Johannes Wirz
Page 132, 135, 138, 140, 146: Mellifera e. V.
Page 165 top: Carmen Diessner
All other photos: Norbert Poeplau
Illustrations, pages 37, 49, 80, 91, 113: Karin Bauer. www.karin-bauer.com

Original typesetting and design: Die Werkstatt-Medien Produktion GmbH,
Göttingen. www.werkstatt-produktion.de.

Johannes Wirz and Norbert Poeplau

Keeping Bees Simply and Respectfully

Apiculture with the Golden Hive

Mellifera Association's 'Einraumbeute'

Preface

Dear Readers

My wish has come true faster than anticipated. Some of you may know the comic book series "Lucky Luke – the man who shoots faster than his shadow". Before I had put out my neck for the search of a translator Martin Kunz, a friend and member of the board of IBRA, announced in a mail that he was willing to translate this book. No sooner said than done. To my delight David Heaf, another beekeeper friend and author whom I have known for many years from former projects, was willing to edit the translation. A heart felt thank you to both of them!

There are a number of reasons for publishing this book in English. I would like to outline some of them. The Golden Hive has already found its way to beekeepers in the UK and the USA, but detailed instructions for managing it have been missing. We at Mellifera have been teaching the basics of bee-friendly beekeeping (alas the expression "wesensgemäß" is missing in the English vocabulary) to hundreds of bee aficionados. The request for practical details by beginners has continued unabated. How do we deal with natural and pre-emptive swarms? How do we manage casts? What are the preconditions for perfect natural comb construction etc.? It goes without saying that most of what is outlined in this book applies to all hiving systems in general.

Although we deal in depth with different methods to cope with the varroa mite, we also show examples of beekeepers – like David Heaf himself – who have achieved a bee husbandry without any treatments whatsoever. Although tremendous losses of colonies without treatments are the rule, such examples spread prudent optimism.

A short chapter is dedicated to beekeeping in schools and kindergartens. It is amazing to see how kids relate to bees. Beside the magic, they learn to understand diligence and the interconnectedness of nature. We trust that Mahatma Gandhi's words "you must be the change you want to see in the world" applies best to children who eventually may remember and carry their past impressions and feelings into their adult life.

For me, bees are the masters of abundance, trust, respect and love – four attitudes for a bright future. I deeply hope that you will understand why.

Johannes Wirz, spring 2021

Contents

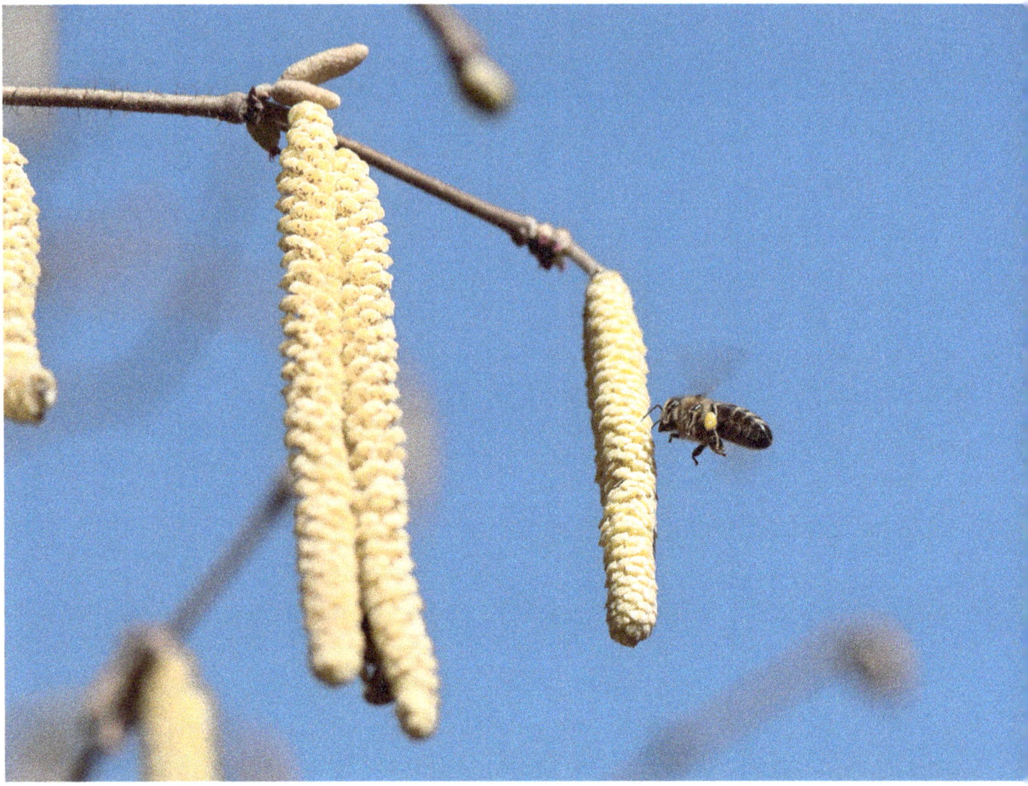

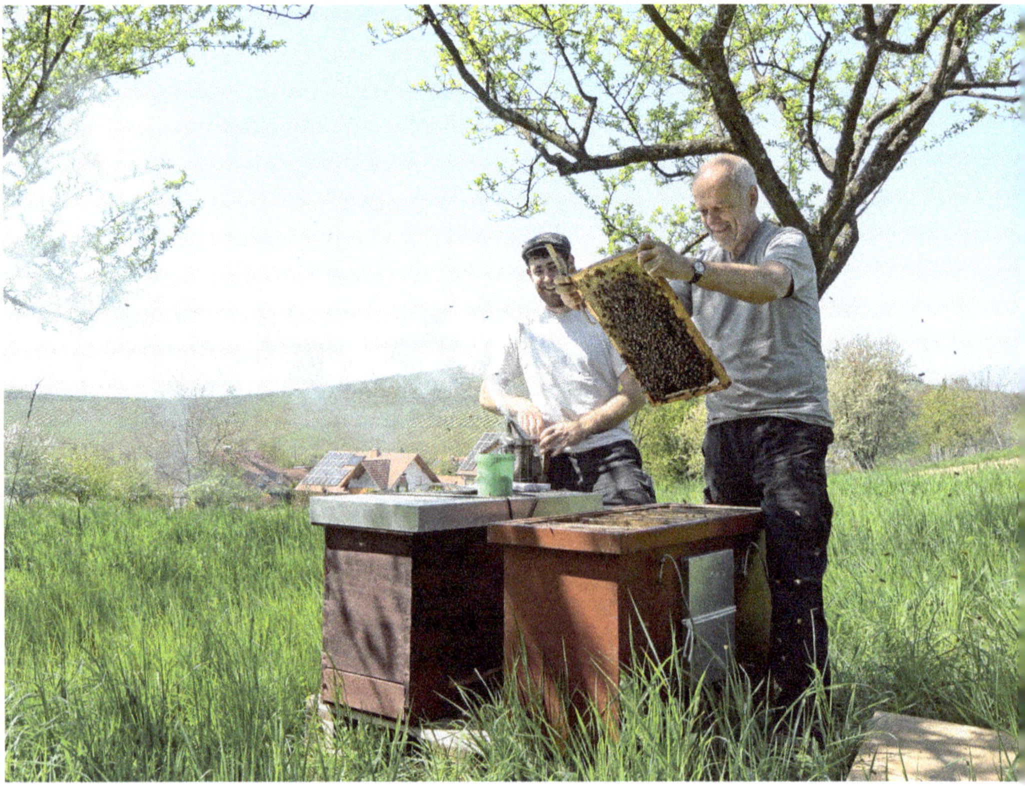

Learning to Understand Bees

Keeping bees in Mellifera's Golden Hive is pure joy – and we hope to convince you of this. Tall vertical frames all on the same level provide optimum conditions for a bee colony to develop naturally. This includes ensuring that the bees have at all times sufficient honey stores for their own requirements. If the beekeeper has to intervene, this can be done without stressing the beekeeper or the bees. In addition, handling bees in this type of hive is simple, and easy on the back.

Although this book is primarily addressed to beekeepers managing their colonies in Mellifera's Golden Hive, it also aims to present our 'bee friendly' beekeeping methods which accord with the bee's inner nature as part of the wider context of 'species appropriate', 'organic' and 'natural' beekeeping methods. In what follows we will call this way of beekeeping biodynamic beekeeping, an expression coined by Matthias Thun. Three characteristics make this approach unique: firstly, our colonies are allowed to swarm, i.e. to carry out their natural and exclusive behaviour of procreation. Secondly, combs are built by the bees naturally, i.e. without foundation. Thirdly, artificial queen breeding is rejected in favour of natural mating and local adaptation. In addition, each colony is regarded as a unique individual. This rules out treating bees like parts of a construction kit.

Anyone keeping bees in a conventional manner may have difficulty following some of these basic principles. But we know from experience that sustainable methods bring beekeepers and their colonies closer together – a closeness and intimacy with the bees are core features of our method.

Living together with our bees enables us to learn to see their needs and requirements, and to understand their way of life.

In our opinion, biodynamic beekeeping is the most bee-friendly way of managing bees – but there is more to it than that. Mahatma Gandhi once stated: "You must be the change you wish to see in the world". This means that we want to be more than just good beekeepers, we want to learn and adopt the 'principles' of this impressive living organism: bees live in and from abundance, which is characterized by their unconditional sharing. Unbiased trust, never control, is the quality of their shared living and working. And in special circumstances, like the swarm's search for an ideal home necessary for its survival, they demonstrate a high level of diligence and respect. We would also like to see all these virtues at work in both the closer and the broader social contexts.

This book has arisen against the background of numerous workshops which took place at Fischermühle, Mellifera's teaching apiary. Even simple tasks are easier to learn, when one can observe them in the real word. And we all remember moments of despair from our early beekeeping years, when we stood in front of an open hive and saw events and processes we could not understand, things which became clear when we could discuss them with like-minded people. Here too, the similarities between bees and humans come to light: many things are easier to deal with when tackled jointly.

For a beekeeper who is concerned about the wellbeing of his or her bees, colonies again and again signal how we can help them. It is our hope that with an introduction to the fascinating life of the honey gatherers, we can facilitate an entry into biodynamic beekeeping. Hence, to begin with, we would like to invite you to take a first look into our a horizontal hive, i.e. the Golden Hive. In what follows we would like to accompany you with practical advice through the beekeeping year..

∾ Advantages of the Golden Hive at a glance

In the Golden Hive both intensive and extensive beekeeping are possible. This hive type has many advantages for both bees and beekeepers:

- ▶ The bee colony develops in a single cavity and is not split into parts.
- ▶ Combs are built in a natural manner without foundation and in most instances sufficient honey is available for the bees' needs during summer.
- ▶ Both the brood nest and the storage area can easily be expanded, without disturbing the colony.
- ▶ When inspecting the brood nest there are no heavy honey supers to lift.
- ▶ The working height of the hive can be individually adjusted at the time of the initial placement and allows for backstrain-free beekeeping.

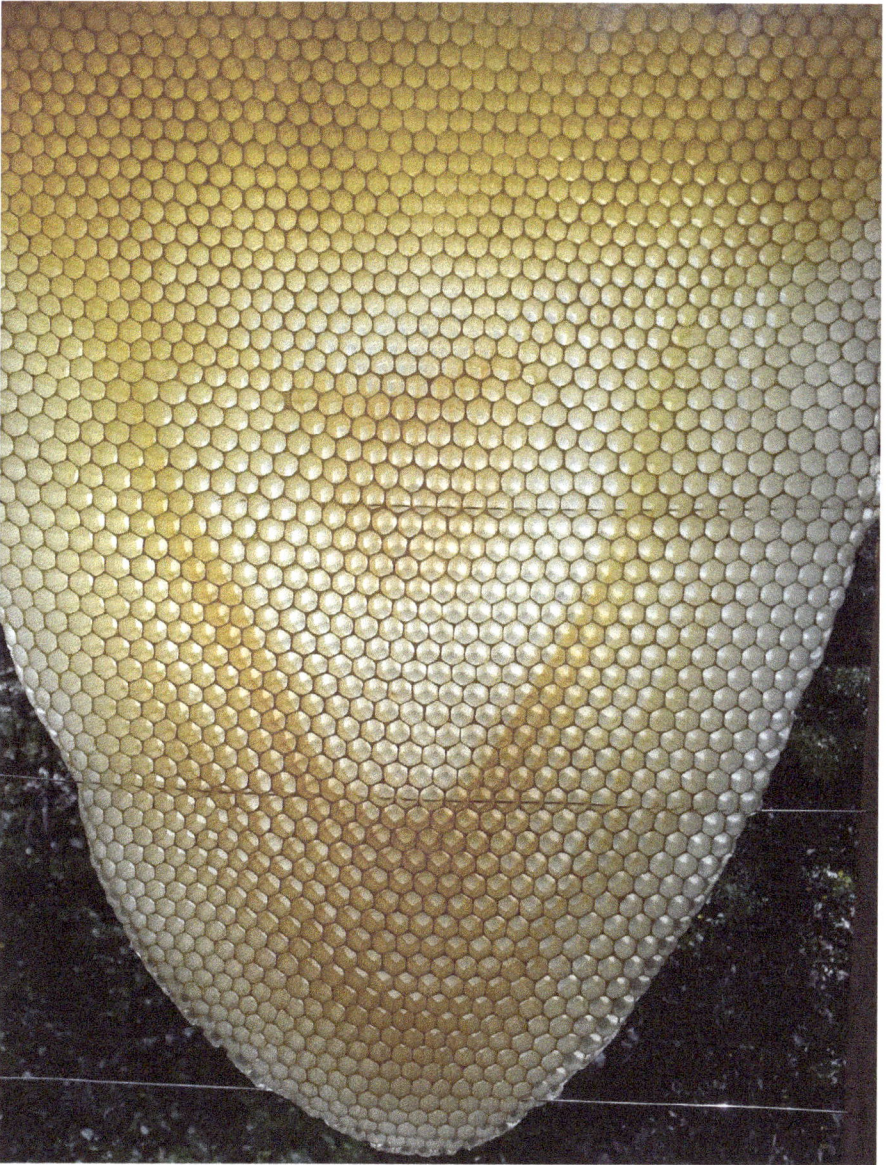

When bees build combs, they do so in the shape of a hanging chain. This freshly built specimen clearly shows the successive sections of construction. Horizontal wires provide stability for the combs and are perfectly integrated.

First Encounter With Our Honey Bees

On a beautiful, warm afternoon at the end of March, on our way to the apiary, we notice a big solitary bumblebee hovering around close to the surface among the sparse vegetation. She is a queen, searching at the end of hibernation for suitable cavity in which to start a colony – something like an empty mouse hole will be fine. A few days later, once she has found a suitable place, she will only be seen visiting blossoms.

Above our heads in a flowering willow tree we hear a quiet, manifold humming. Hundreds of bees are collecting the abundance of gold yellow pollen. The deprivations of winter are history, the flowering of the willows marks the beginning of the new bee season. The contrast could not be starker: Here droves of animals, there a solitary insects. We guess already that a single bee – in contrast to a bumblebee – can only be understood in the context of her colony.

At the apiary there is plenty of activity. We pause for a moment next to the entrance hole of the hive, which we plan to open in an instant. Bees are flying purposefully in various directions into the wider surroundings. Many homecomers carry yellow pollen pants, others return without a visible harvest, but their flight seems to be clumsy. They land heavily onto the alighting board, often they even bump onto the ground in front of the hive. They carry sweet nectar or water, which is of vital importance in order to liquidize the honey for blending the royal jelly. Other bees leave the hive, but turn around almost immediately, and perform swinging flights. These are young bees on orientation flights: Before they start to forage, they need to get to know and memorize the surroundings of their hive, in order to find their way home safely after often long distances travelled to flowers and blossoms. The activity at the entrance hole radiates calm. The bees and their actions fit into a whole, which remains invisible, since it is hidden inside the hive.

Tracing Clues in the Debris

After we have observed the activities at the entrance hole for a while, we pull out the floor board, i.e. the sheet which has been fitted underneath the mesh floor. This sheet shows a space-time image of everything that has dropped down from between the combs above. We can see, how many of the inter-comb spaces are occupied by the colony and therefore get an idea about its size. Fine dark brown particles on the sheet indicate remnants of cell cappings that emerging young bees chewed away when they freed themselves from their cradle. Lighter coloured bits are from honey-stores cappings, which have been opened and emptied. And finally we can also observe some pollen loads, which have been dropped in the process of storage. There are also tiny

The debris on the bottom board reveals the activities of the bees: Yellow pollen has been collected, but lost while walking across the grated floor. This is particularly the case near the entrance hole to the right. Caps opened by hatching bees make brownish crumbs. Wax scales appear as white dots

The building of combs only works together. Bees clinging to each other like a curtain sweat scales of wax and shape them into cells with fascinating precision.

transparent scales of freshly sweated wax – an indication that the colony is building new combs. In some instances, a different type of small black particle is visible indicating the presence of larvae of the wax moth. There is no need to worry, because they often co-exist with live colonies. Their natural task is to return old wax from abandoned combs to the natural cycle. Conspicuous damage by the moth in hives is a sign of sick and weak bee colonies.

Constructing a comb is a communal task. Whilst it is true that an individual bee can sweat wax scales from the pairs of wax glands on her abdomen, without the warmth generated by the colony these scales would quickly cool off and loose the malleability crucial for the construction of new comb. This is the reason why there are always clusters of bees sitting on freshly constructed comb, producing the necessary temperature. The first cells and combs are built completely hidden away in the bee cluster and therefore almost invisible. Only after the combs have grown in size and most of the bees are busy with the care of the brood or the processing of honey in between the freshly built comb become visible. Then they seem almost to grow out from the bulk of bees forming a mobile curtain of bee bodies linked like chains. In these comb construction chains the bees forming a 'curtain' are exclusively occupied with sweating the wax scales, adding and moulding them to the many emerging new cells at the bottom of the growing comb. Within these chains, the builder bees are always hanging plumb-vertical. Therefore without exception combs are always built perfectly vertically.

There may also be a few varroa mites on the bottom board, either lifeless or moving sluggishly. We will keep a close eye on them. This parasitic arthropod is making life very difficult for European beekeepers. Only in very few cases and under very special circumstances can colonies coexist with varroa mites. Beekeepers in Germany are obliged by law to treat against these mites, at the very latest when certain thresholds are reached. The number of dead mites falling naturally to the floor board provide an indication of their total number in the colony. More on this on page 115.

The debris on the bottom board is a record of the developing bee colony. It tells of its size, its activities, its biography. If it is not well structured, it means that the colony is not living in a compact cluster – either because there are not enough bees, or it is queenless, or its queen is no longer capable of ensuring the integrity of the colony. If there are big pieces of wax on the bottom board in a chaotic manner, then there may be open or silent honey robbery going on. A stickiness around the entrance hole would confirm such a suspicion. Broken antennae indicate hygienic behaviour. Bees remove diseased or mite damaged bees before they hatch.

Worker bees looking up from the space between two combs. The edges of the frames are covered by propolis.

At Eye Level With the Bee Colony

After these first important observations, we lift the roof cover and the insulation board. We can then already tell by lightly touching the wax cloth that is in place for additional protection of the colony, where the brood nest is located, as this is kept at a temperature of 35-36°C, i.e. higher than the ambient one. Tracing the warm spot with our fingers, we will notice that the brood nest is built in a circle. The diameter of the circle of warmth is largest on the combs at the centre and it diminishes towards the combs at the edge. Where the honeycombs are situated no warmth is felt, because the bees do not warm them. There is hardly any difference between ambient temperature and that of the honeycombs.

We carefully pull back the wax cloth. An indescribable fragrance emanates from the hive. Pollen and nectar, of course, but also the brood nest contribute to a flavour which we have been anticipating all winter long. A pleasant fra-

grance is an indication that the colony is well and healthy! No single bee on its own is able to produce this scent, only the colony as a whole can do so.

The Golden Hive is ideally suited for inspections. Beginning at one end we use our hive tool to loosen the frames of combs fixed tightly to the hive by propolis. The bees have been busy closing every single small hole and cleft and 'cementing' the frames to the box over winter. (Propolis is a blend of resin collected by the bees from the buds of deciduous trees mixed with secretions they produce from special glands).

The frames on the edges should still be filled with stored food. As a rule of thumb, there should still be two thirds of the winter food available in March. In the cold period, when there is no brood, a bee colony requires extremely little food. Only when the brood nest is established again do worker bees raise the temperature to 36°C by their flight muscles.. The bigger the brood nest grows, the more stores are 'burnt' to provide the required energy.

Roof cover, insulation board, and wax cloth have been removed. The frames stuck with propolis, loosened with the hive tool, can now individually be pulled up for inspection.

Combs at the centre often have only a thin semi-circular rim of honey stores at the top of the frame, while the cells below are empty and have been cleaned until they shine. Soon the queen will lay eggs in these cells. Checking frames closer towards the centre we will encounter the first comb with brood: a small circle of cells with eggs, which are shaped like tiny grains of rice. On the adjoining frame there will be a bigger area in the centre with larvae, floating in royal jelly.

At the very centre of the brood nest we will encounter comb with capped cells. Behind their lids, which, unlike the lids on the honey stores, are permeable to air and moisture, white larvae have pupated. In the pupa, over a period of 12 days the mysterious transformation into a bee with head, thorax, and abdomen takes place.

The area of covered brood is surrounded by cells with large larvae which become smaller and smaller towards the edge of the brood area, until there are eggs from which no larvae have hatched yet. The circular structure of the brood nest is the result of the 'path' taken by the queen when she lays her eggs following a spiral pattern from cell to cell (see illustration on page 20).

Adjacent to the central comb the circular comb areas full of brood and eggs get smaller in a mirror image pattern. At the beginning of the 20th century, Ferdinand Gerstung, a pastor in Weimar, was the first to document the egg laying pattern. When he put a swarm into a hive with seven empty combs the queen began laying eggs in a small area in the centre of the middle comb. The next day she crossed to the adjacent comb on the right and again laid eggs in a circle again. Then she returned to the central comb and expanded the nest by laying more eggs in a ring around the original circle. From there she moved to the empty comb on the other side of the central one, then back to the centre, then back to the next one to the right. From there she jumped to the next empty comb on the right, after which she made her way back to the other side. Finally, with an increasing pendulum movement she established a round, spherical (in three dimensions) brood nest across those combs. On day 24 she returned to the centre of the central comb, where the first bees had already emerged and cells had been cleaned ready for the queen to start the second generation of her daughters.

Around the brood nest, pollen is stored in a circle on every comb followed by honey in a semi-circle. What can we learn from this structure? We have already felt the round shape of heat on the wax cloth when we touched it with our hands. If we put the two observations together it becomes obvious that the brood nest, and hence the whole way in which the colony is organized, is shaped like a sphere – the archetypical shape of a bee colony.

We can now put the many details together in our minds and reconstruct the whole. Even though we always see just individual bees – as well as eggs, larvae in various stages of development, and covered brood – it becomes clear that all activity in the hive is geared towards the colony as an entirety. Whenever we lift a frame, we are destroying its integrity for a brief moment. It becomes easy to understand why Ferdinand Gerstung called the removal of frames a surgical intervention (see also page 28 on this topic). Observations at the hive entrance, a check of the pattern of debris on the floor board, the feel of warmth on the wax cloth, and the inspection of individual combs are all snapshots of a complete whole: the bee colony. But only in our own minds, not in those individual appearances, does the hidden life of the Bien become real for us.

Ferdinand Gerstung was the first to document the egg laying walk of a queen. Within 24 days she establishes the spherical brood nest on seven combs in a spiral like movement, moving back and forth one comb to the next in a growing pendulum like movement.

Clearly visible on this picture of a comb that is part of the brood nest covered by heating bees: Plenty of pollen (bright yellow), and honey in capped cells (top left corner). The bees warm the brood nest and supply the larvae with royal jelly during the first three days to which they subsequently add pollen and nectar to it when the larva get older (below).

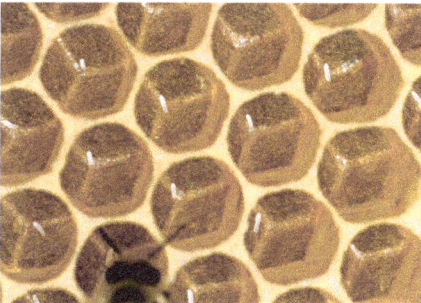

21

Bee friendly bee keeping – a question of awareness

This short introduction gives an initial impression of the way bee-friendly bee-keepers approach the Bien. Every time they meet the Bien they feel that they are confronted with an organism in its own right, and not one that is simply the sum of the individual bees. In contrast to the encounter of a cow or a dog, the bee colony as a whole can only be fully comprehended through the inner experience and recognition of the beekeeper. As in encounters with the cow or the dog, they meet the Bien[1] with esteem and respect, knowing beyond any doubt they will never fully comprehend the complexity of the organism.

The comparison with the relationship to our farm animals and pets can be taken even further. The process of domestication turned the wolf, which roamed the forests in packs far away from any human influence, into a dog. He is affectionate, faithful, even submissive, and cannot to be close enough to us. The cow too was originally living in forests, at times in large herds. Nowadays they live in stables, are milked and taken daily to pasture. Under the care of man it gives up parts of its original nature. Those who keep animals in a nature-appropriate way are aware of this trade-off. At the same time they try to ensure that the animals can satisfy their most basic or even 'essential' needs and live their lives accordingly. Anet Spengler Neff of the Research Institute of Organic Agriculture (FiBL) in Switzerland, has shown that the cow, with its many stomachs, is a typical grass feeder, i.e. an animal that ideally converts roughage into energy. Feed concentrate, which is often used to increase the milk yield, is alien to its inner nature.

The situation with bees is similar. They too have essential needs that should be taken into consideration. Feral bee colonies are solitary organisms, they

---

1 'Bien' is a German word that is becoming accepted terminology when referring to a bee colony as a single organism. Horst Kornberger (author of 'Global Hive: What The Bee Crisis Teaches Us About Building a Sustainable World') has suggested 'essential bee' as an English equivalent – which in my opinion does not catch the concept of a 'composite unit' as well as 'Bien' does. (mk, translator).

Multiple options are possible for encounters with the Bien at the training apiary at the Fischermühle/Germany. Visible in the foreground are bees flying to and from a log hive. In the background there is a Sun Hive a creation by the sculptor Günter Mancke.

avoid neighbours within one square kilometre. And they live fairly high up (six to eleven meters above the ground), in tree cavities. Today in apiaries under human care, groups of colonies are kept pretty much at ground level. Their species-appropriate behaviour is collecting pollen and nectar. Keeping this in mind, sugar should only be fed in an emergency. (More on other aspects on page 28). Compared to the situation in nature, any animal husbandry by represents a compromise. This is not an excuse – on the contrary. In every relationship between beings, and this is true for relationships between humans too, all those involved give up some of their original individuality, in order to create space for something new. An additional feature of such relationships is their permanent, ongoing evolution – characterized by respect and reverence.

Origins and evolution of the relationship between bees and man

There are quite a few people who believe that bee-friendly beekeeping is an invention of the 20th century. In a literal sense this may be correct, but in a broader sense this already applies to the origins of keeping bees. Species-appropriate beekeeping was the first type of interaction between humans and bee colonies. The oldest depiction, a picture from the Cueva de la Arañas in the Spanish province of Valencia, leaves no doubts about this.

A human person clings to a fragile rope ladder, at eye level with the bee colony, returning bees are drawn unnaturally large. The exact age of the cave drawing is unknown, but it is at least 8,000 possibly even 12,000 years old – and its aesthetics are deeply moving. On a superficial level we are shown the harvesting of honey in a way that even today is still practised in some countries.

Like other representations of this era showing animals and humans, for example in the caves of Lascaux in Southern France, the drawing in the Cueva de la Arañas (according to anthropologists) show an awareness enabling humans to experience spirits as well as spiritual qualities in all animals, plants, rocks, and mountains. The French ethnologist Claude Lévy-Strauss said that animals are good not only for eating, but also for thinking. This statement points to a mythology in which animals and plants were not only important for human sustenance, but also were of equal symbolic and spiritual importance. In those times, in every phenomenon, humans encountered at the same time the being which became visible in this particular manifestation and which had produced it. We may describe this pristine type of relationship as horizontal culture.

In the transition of the nomadic life of hunters and gatherers to sedantariness and the beginnings of agriculture in Old Persia, this relationship underwent a radical change. From that time onwards, spirits and gods no longer lived among humans, but instead in human imagination they said from the heavens above: look after the earth! This separation of heaven and earth,

mind and matter, marked the beginning of vertical culture. The famous drawing of the Egyptian beekeeping priest demonstrates this change.

The picture of the kneeling beekeeping priest is approximately 5,000 years old. It shows the beekeeper priest in prayer with his hands pointing heavenwards. The bees are still shown in a supernatural size, a sign of their importance. The implements in front of the kneeling priest are pipes made from clay or Nile mud, which had been placed in an urban apiary – which shows just how ancient is the concept of 'urban beekeeping'.

We owe not only a settled lifestyle and agriculture to vertical culture. Both are preconditions for the evolution of the culture of modern man. Without it there would be no sciences, no religion with its cathedrals, no art, no symphonies.

But this progress comes at a price. The separation between heaven and earth, spirit and nature, God and man, came to an initial climax during the age of Enlightenment in the 17th century. The divine in nature, as well as the one in bee colonies had lost its place in human consciousness. From then on, to understand meant to observe details, to dissect and analyse. It was no coincidence that the first insect to be dissected and viewed under a microscope happened to be a honeybee.

In particular the invention of hives with movable frames of combs in the 19th century made it possible to observe and unravel some of the (biological) secrets of a bee colony. In a skep, a cavity in a tree, or a log hive the comb was firmly attached to the walls of the colony's home. In order to study the organisation within a colony it was necessary to destroy its unity.

In addition to the moveable frame hive two further groundbreaking inventions characterize modern beekeeping: the introduction and use of wax foundation with fixed cell sized embossed, which the bees quickly draw out into full comb, and artificial queen breeding. This enables the beekeeper to raise as

Combs were firmly attached to the walls of bee homes in skeps such as the 'Karnitz'-Ring (left) and in the 'Allemannischer Rumpf' (right).

many queens as he wants, and he further may control the mating process by bringing virgin queens to a mating site with selected drone colonies. Artificial insemination takes breeding a step further, at times making use of the sperm from a single drone only.

However great were these scientific discoveries and insights, and we will later get to know some of them, equally great was the temptation to make use of them for interfering with and manipulating bee colonies as we see fit. The colony can be taken apart into manageable parts like a construction kit: comb with brood and stores, bees, queen and drones can be moved around at will and rearranged in whichever way it is the mechanisation of beekeeping par excellence.

The rediscovery of the inner nature of bees

At the beginning of the 20th century came the first thoughts and admonishments of those who, with deep understanding and equal passion, started promoting an 'organic' mode of beekeeping closely in line with the inner nature of the colonies. Two personalities made a huge impact in the context of bee-friendly beekeeping: Ferdinand Gerstung (1860-1925) and Rudolf Steiner (1861-1925). Both lived for a time in Weimar, but their paths never crossed.

Gerstung was a pastor, but saw himself as a scientist. Bees were his focus of research, since he kept hives himself and was of the opinion that it was possible to dissect the colony as organism – unlike that of other animals – without drawing blood or killing the object of the study. In 1920 the University of Jena awarded him a honorary doctorate for both his organic conception of the bee colony, which, like a few others before him, he called the 'Bien', and his general achievements in bee research.

He researched multiple individual aspects of the bee colony, and described the individual worker bees, drones, and queen as 'organs' of the Bien. A condensed summary of Gerstung's observations reads: "The Bien is an organism, which consists of the harmonic and functional co-operation between all its parts and organs, and in which each part implies the existence of the whole as origin and agency of its own existence".

He described the spherical shape of the colony that is clearly manifested in the brood nest organisation and the construction of comb. He strongly emphasized the need to respect and maintain the Gestalt of the colony. He called the forager bees who collect pollen, nectar, and water, the 'arms of the Bien'. His records show that he considered the bee colony to be a dynamic entity not only in its physiological, but also in its anatomic structure. In the winter cluster it will form a round sphere of the size of a foot- or a handball, in summer it will grow into a giant that is spreading out into the landscape over a number of square kilometres.

Following a request by a beekeeper, Rudolf Steiner, the founder of anthroposophy, gave nine lectures on bees to the craftsmen engaged in the construction of the Goetheanum[2] on a hill near the Swiss village of Dornach. His

---

2 https://en.wikipedia.org/wiki/Goetheanum

remarks were unusual and opened up new perspectives on the life of a colony, which he described as a 'head without a skull' open to all directions. This description recalls the ball-shape of the Bien as described by Gerstung. Later the United States bee researcher Thomas Seeley put it similarly, defining a swarm of bees as an open brain.

According to Steiner the queen is the organ of a colony's inner unity. This is being threatened by artificial queen breeding. He called the deliberate exchange of queens – common practice among many conventional beekeepers – the 'mechanisation' of beekeeping and argued that it may be prone to destroy the inner cohesion of colonies. He even prophesized that "bee breeders can celebrate the exceptional upswing noticeable in recent bee breeding, but this joy will not last 100 years." When in the winter of 2006/2007 beekeepers in the USA lost up to 90% of their hives, a few of the major newspapers in the United States pointed to the prophecy of the 'Austrian philosopher'.

According to Steiner, the comb is the skeleton of the bee colony. He describes wax in relation to its function as a composition of bone, blood, and muscle tissue: just as in a mammalian embryo, the bones develop surrounded by cells, so also the combs when a swarm initiates their construction are invisible. The cluster of bees, which is hanging from the ceiling of the new home, construct the first combs in their midst. Only when they combs have grown to a certain size can they be observed growing out of the bee cluster. The comb is the biggest organ of the colony (in total approximately 1.5-2 sqm), this skeleton provides the vital support of the colony. Similar to the growth of the blood cells in the marrow of the bones of mammals, the comb is the cradle of the worker brood, which Steiner called the blood cells of the Bien. Muscle tissue is always associated with motion – and this can be observed in comb, too. Naturally built combs are only fixed to the frames at a limited number of points. Thus, they easily start to vibrate when forager bees perform round and waggle dancing. These vibrations can be sensed and interpreted by their sisters on the same comb.

Seeing a swarm issuing from a hive leaves a deep, lasting impression on all observers. Initially bees tumble out of the hive like a small waterfall and then fly in a chaotic manner hither and thither, merging into a dancing swarm cloud, and finally settle in a cluster onto a branch of a bush or tree nearby. From a spiritual-scientific perspective, Steiner calls the issue of a swarm the 'near death' of the colony. The swarm leaves behind everything: shelter, i.e. the hive, the comb-skeleton, the stored food, and the brood, which are so to speak the 'stem cells' of the colony, since they constitute its rejuvenating.

The 're-birth' of the colony takes place in two steps: the first is the swarm cloud of bees clustering around a branch. The second happens when the swarm looks for a new home, or is collected by a beekeeper and housed in a hive box. Only then is it possible for the young colony to develop and grow.

∾ The Bien

A bee colony, as a collective organism, is much more closely related to mammals than to other insects. The following characteristics stand out:

- ▶ capability to precisely regulate its own temperature,
- ▶ differentiated 'organs' such as cleaner bees, nurse bees, construction bees, guard bees, foraging bees and heating bees. The last task is executed irrespective of the bee's age,
- ▶ immune system to protect against diseases at the level of both individual bee and colony,
- ▶ complex system of information with a symbolic language, consisting of some four different dances, but also scents (pheromones) of the queen, as well as from pollen and nectar,
- ▶ feeding of the brood with secretions from special glands.

Confirmations from current bee science

In 1995, after many years of development work, Thomas Radetzki, together with other pioneers of the Mellifera association, drew on the indications of Rudolf Steiner on the inner nature of bees, and defined the standards for biodynamic beekeeping. They explicitly include that colony reproduction is permitted only by the process of swarming, that the combs for the brood nest have to be built naturally by the bees themselves as an undivided unity without interruption by bottom or top bars of frames, and that the artificial breeding of queens is not permitted.

These regulations used to be ridiculed and rejected for various reasons by conventional beekeepers, e.g. the limitation of colony propagation to the time of swarming from Mai to June, or because the colonies have to expend a lot of energy on comb construction. The production of one kilogram of wax requires approximately six kilograms of honey. However, it does not necessarily follow that this weight will be missing from the honey harvest. If the nectar flow is good, even young colonies can produce a decent honey harvest.

Another prejudice claims that docile and productive colonies can result only from targeted breeding. This opinion is incorrect. In bee-friendly managed operations, many beekeepers work all year long without protective clothing or veils. Docility can also result from refraining from trying to influence the selection of drones for the mating of the queen, and instead allowing her to go on mating flights from her own hive. Furthermore, studies at the bee research institute of Liebefeld/Switzerland have shown that the honey harvest does not depend on queens bred for high performance, but instead is reliant on sufficient forage in the vicinity of the beehive.

Results from current bee research also show, that the principles of bee-friendly beekeeping – propagating colonies by swarming, natural construction of combs, avoiding artificial breeding – give a number of further advantages.

In conventional beekeeping, the suppression of swarming by repeated removal of queen cells is standard practice. It is even claimed that swarms only occur from hives kept by 'poor', i.e. inexperienced beekeepers.

Swarming and bee health

A common accusation levelled against beekeepers working with the process of swarming, criticises the production of 'disease spreaders'. Swarms are believed to look for a new home somewhere and go on to develop diseases because lack if disease monitoring, which in turn would jeopardise the colonies of other beekeepers. This opinion is plainly wrong. It is assumed that the many swarms that unintentionally leave their hives – in Switzerland this number is estimated to be 30,000 – actually occur in apiaries in which beekeepers attempt swarm suppression. It is well known that if only a single queen cell is overlooked in a colony, it is certain that a swarm will leave the hive.

The importance of swarming for the health of colonies must be seen as a basic tenet of bee health, which has evolved and improved again and again over the whole evolutionary process lasting millions of years. Interruption of the breeding cycle both for the prime swarm and the casts results in a significant reduction of brood-borne diseases. This has been shown for European and American foulbrood. In addition, beekeepers working with the swarming process know well that the load of varroa mites is reduced substantially (more on this on page 109). A remedial action to rescue individual bee colonies from acute foulbrood and European foulbrood infections, is the building of artificial swarms resulting in the interruption of the brood cycle (see page 131 on this).

The US-American bee researcher Thomas Seeley has demonstrated impressively the benefits that swarming brings for the bee health: 12 colonies each were housed in small hives with a volume of only 40 litres, and in large hives with 160 litres. They were then analysed in the following year. In the small hives 80 per cent of the colonies had swarmed, in the large hives the number was below 20 percent. The number of phoretic mites on bees was three times less in the 'small hive colonies' compared to the 'large hive colonies'. The former showed no damage by the deformed wing virus, whereas seven of the twelve colonies in large hives had bees with crippled wings. And with respect to winter colony losses, there were only four colonies lost out of twelve from the 40 litre hives, in contrast to colonies in the 160 litres hives where ten out of 12 colonies did not make it to the following spring. Not least for these reasons, scientists at German bee research institutes have developed biological methods against the mites, all of which include an artificial interruption of the brood cycle. (Details of these methods on page 126).

Natural combs and bee health

For keeping a colony healthy, sweating wax and building natural comb are, after swarming, the second most important hygienic measures. Positive effects on American foulbrood and European foulbrood have been demonstrated, the latter being of particular concern in Switzerland. For both the recommendation is to make artificial swarms with infected colonies. Following that the swarms are put onto empty frames or frames with fresh foundations. This method helps to successfully cope with these fatal diseases. Nevertheless, even today, colonies are still being destroyed even in cases of minor infections.

The construction of natural comb is almost unheard of among conventional beekeepers. The following story is an example for this: a few years ago, with the co-operation of Thomas Radetzki, two colonies in Golden Hives were placed in the park of the Beyeler Art Museum in Basel. Ernst Beyeler, the founder and owner of the famous Basel based gallery, pointed to his name, which, when translated, means 'beekeeper' and commented: "I collect art, bees collect flowers". Hence it made sense to him to place beehives in the park. And because we told him that in the old days honey had been so precious, that it was never sold, only passed on as a gift, until today honey is not sold in the museum shop, but only given as a present to sponsors and friends.

After placing the hives we invited the bee inspector from the city of Basel for a tour. This gentleman, who was more than 60 years old, had never seen natural comb construction before, because he had always used frames with ready-made foundation. He was surprised and delighted by the perfectly built combs.

Since the invention of foundations more than 150 years ago, most colonies are raised on such sheets of wax. They are produced in an industrial

Two Golden Hives in the park of the Art Museum of the Beyeler Foundation in Basel.

process and rolled or moulded to result in a fixed cell size – mostly 5.4 mm. On such a foundation the bees have merely to draw the cell walls. The eggs laid by the queen into such cells develop into worker bees. Large cells with a diameter of 6.9 mm, in which unfertilized eggs develop into drones, are nearly non-existent – a method to control drone brood. Only in later years, empty frames are added, which are readily filled with drone cells by the bees. These combs then serve as varroa traps, and once the brood in them is capped, they are removed from the colony and destroyed. As a result of this practice, the stubborn prejudice persists that bees are only capable of building drone cells, and does so despite that fact that over millions of years the bees have developed and perfected the capacity to build natural combs.

But we can also detect a change of attitude with respect to natural comb building. Not least because foundation has been sold with the addition of stearin and paraffin, more and more beekeepers seem to remember that bee colonies can produce and organize comb building by themselves.

Natural comb has another important advantage: in most instances, the comb shows less contamination by fat soluble, synthetic varroa treatment chemicals and pesticides, compared with combs on foundation, which has been reused for innumerable cycles and without knowledge of the provenance of the wax. Whoever has taken a look at the websites of agricultural agencies and ministries, will have found 100s of products permitted for use in conventional agriculture against pests, fungi, and parasites. Most of these products are fat soluble and therefore will accumulate in wax. In a closed wax cycle, a normal procedure in conventional beekeeping operations, these bee toxins accumulate over the years and find their way back into the colonies. Today it is possible to detect not only the infamous bee damaging neonicotinoids but even DDT, the application of which has been banned in German agriculture and forestry since 1972. All these compounds are a serious stress factor for brood and bees alike.

Natural mating and bee health

Nowadays, the avoidance of artificial queen breeding by bee-friendly bee-keepers is also no longer considered wrong. Keeping bees in an organic system, swarming and relying solely on natural matings of the queen are integral parts of colony management. Scientific research has well documented two beneficial consequences of this: a high genetic diversity among the colonies, which is promoted by multiple matings at regional drone congregation areas, and an adaptation of the colonies to their respective locations. Most certainly characteristics like honey productivity and docile behaviour do have genetic components, but it is questionable whether it is possible to actually improve such traits by targeted breeding. Like many other characteristics these too are strongly influenced by the style of beekeeping, management, and location. Anyone keeping bees knows that limited hive space at the time of the nectar flow will increase the tendency for swarming, whereas providing more space will reduce it. The volume of the honey harvest is first and foremost the result of the location of a colony and the seasonal conditions. A comparison in Switzerland between colonies with high performance queens and others, that had not been specially selected, showed no significant differences in the colony size and honey yield when they were placed in the same apiary. But when, as part of the same experiment, these colonies were taken to different locations, the differences were statistically significantly.

Another example is the issue of docility. A few years ago, 50 students from Hohenheim University near Stuttgart visited our teaching apiary at Fischermühle. Over the course of three hours some eight to 10 hives were opened and inspected. No one was wearing a veil and no one was stung. The professor in charge of the group wanted to know how we had managed to breed this level of docility, which he had never before seen elsewhere. He could not believe it had been maintained and improved despite(!) allowing for only natural mating.

On the other hand, many beekeepers have also experienced that docile colonies can become quite aggressive, and conversely, that a very aggressive colony can come out of the winter break and into the new bee season totally peacefully and docile.

Even with respect to bee diseases, where efforts have been going on for years with aim to breed and select colonies with resistance or tolerance, no positive

The three bee beings

Worker   Drone   Queen

results worth mentioning have so far been achieved. The reason for this lies in a peculiarity of the honeybee, namely multiple matings or, expressed scientifically, polyandry. On her multiple mating flights – sometimes spread out over a number of days – each queen is inseminated by between 12 and 40 drones. If the weather is good, 100s of drones from all over the region will fly around at drone congregation areas. Since queens on average fly five kilometres, but drones only three kilometres, there is hardly any chance that the queen will mate with a drone from her own colony. The largest possible genetic variety is a strategy for survival, which has been proven successful through evolution – it is species appropriate. During the last couple of years, multiple studies have confirmed the importance of polyandry. Because of an increase in genetic diversity as the result of eight to fifteen fathers – compared to colonies with one father only – the colony strength at the beginning of the winter season and at the beginning of the new season was improved. The brood nests were larger, more combs were constructed and heat regulation was better. An increase in the number of drones that mated with a queen will also result in more intense waggle dances with a concomitantly higher level of recruitment of foragers. In addition, the bees fly to more distant nectar sources and thus improve nectar and honey yield.

Within a colony, too, the importance of multiple matings has been demonstrated: there was an increase in resistance against infectious diseases, and the defence against artificially introduced foulbrood spores was improved.

Because bees visit flowers which are further from the hive, there is an increase in the variety of the microflora and fauna and hence the longevity of pollen storage is improved, a benefit for feeding the brood in times without any incoming pollen.

It is unlikely that targeted breeding programmes could have produced the numerous important and positive characteristics resulting from polyandry.

Genetic variety is also the guarantee for a large pool of behavioural capacities. The Würzburg based bee researcher, Jürgen Tautz, paints a vivid picture of the situation: if the descendants from one drone line were good for foraging, but not good for tasks inside the hive, the colony would suffer to the same extent as if it only had workers good for internal duties. In the first instance brood caring would be poor, in the second the collection of pollen and nectar would be insufficient. Multiple matings ensure a balanced spectrum of all necessary functions and behaviours within a colony and its surroundings. The result is far better than the sum of all individual skills.

Since the loss of genetic diversity as a result of the issues outlined above presents one of the big problems in modern beekeeping, in our opinion the right way forward is to forget about controlled and targeted cross breeding and instead promote the strengthening of bee colonies by allowing natural matings. This gives the bees the added chance to make use of another benefit that has become very important in the disciplines of ecology and evolutionary biology under the label 'adaptation'. All beings, whether plants or animals, will over time and generations adapt to their surroundings. Whether this is the result of pure coincidence, as claimed by neo-Darwinian thinking, or is a purposeful activity, as suggested by epigenetics, does not really matter in this context.

∾ Natural mating and its advantages

Natural mating describes the process where colonies raise their own virgin queens, which then mate on their mating flights at a drone congregation area with twelve to twenty drones on average. Since drones from hundreds of different colonies congregate at these areas, a large genetic variety is guaranteed. But since all drones come from within the same general area as the virgin queens, a natural adaptation of the colonies to their surroundings is assured, too.

Adaptations lead to heightened levels of vitality, and increased levels of resistance. This holds true for bees, too. There is strong evidence for this in Europe. In a number of regions colonies have managed within a short period of time to develop a co-existence with the varroa mite, i.e. to survive in the presence of the mite without treatments.

The importance of the adaptation of colonies to their respective locations has been confirmed under the framework of a large study organized by the international research consortium COLOSS (Prevention of honey bee COlony LOSSes). For a number of years in many European countries 612 colonies were observed, comprising 16 different races at 21 different locations. Colonies which were locally adapted were compared with those made with imported queens. In this big research project no treatment against varroa nor any other disease was done.

Colonies that were locally adapted survived on average 83 days longer, i.e. almost one third longer than the non-locally adapted control colonies. In addition, the colony strength of the former was 3,000 more bees. The honey yield tended to be higher, although without statistical confirmation, and – unexpectedly – the mite load was smaller. The adapted bees were more docile than the non-local colonies, and they also displayed better grooming behaviours. Even though variety and number of pathogens in both groups were similar, the adapted colonies showed better health than those in the control group, demonstrating their improved capacity to deal with the base load of pathogens. One of the conclusions drawn from this study by Ralph Büchler from the bee institute in Kirchhain, Germany, stands out: "Using local populations of honeybees improves the chances of survival, whereas with non local colonies losses must be expected, as was recently observed in many regions. Hence the breeding of local races should be promoted and encouraged for all of the Apis mellifera species."

In our opinion locally adapted bee subspecies develop when queens are mating at drone congregation areas.

All of these examples show that principles of bee-friendly beekeeping improve the health and vitality of colonies, thus enabling the future for beekeeping.

The path to the Golden Hive

Originally honeybees lived in tree cavities in forests from Central Europe to the Ural Mountains. These cavities result from branches breaking off, or when black woodpeckers prepare a home for nesting. In both instances, as soon as water gets into the cavity, fungi start a process of rot. But black woodpeckers only nest where there are ants available, their only prey for feeding their chicks. In turn, since from a colony of feral bees some 200,000 individual bees die per season, they provide welcome food for the ants. Studies have shown that the number and size of ant hills increase if bees are present in the forest – a wise interconnectedness of ecological systems.

What are the properties of cavities selected by feral bee colonies? The US-American bee scientist Thomas Seeley and colleagues, after decades of research, have deciphered criteria that must be fulfilled by potential homes for bee colonies living in the wild:

▸ Ideally, the cavity is three to twelve meters above the ground.
▸ Its optimal volume is between 40 and 60 litres.
▸ The entrance hole is towards the bottom. Its size is 15 cm2, and a round form is better than a square one.
▸ The entrance hole faces towards the South or South East.

In his book 'Honeybee Democracy' Seeley describes in detail the care bees take when checking out at least twelve to fourteen new possible homes, before a swarm takes off. In an open 'decision making process' the bees agree on the optimal choice and ensure via ingenious mechanisms that indeed the entire swarm ends up in the new home. In the natural context, these processes are of existential importance in the truest sense. This is because only if all the conditions are right, to the best possible extent, does the swarm have a chance to survive the coming winter in its new home. Only because strong natural selection has been at work is it possible for us today to rely on this complex and extremely flexible coexistence of bee colonies ensuring pollination on the one hand, and the production of honey and all the other bee substances on the other hand.

Even the very first form of bee husbandry, namely zeidlerei, the management of colonies in tree cavities, excluded some bee-friendly criteria. In the Middle Ages, forest beekeepers were well respected bee husbandmen, who cut into living trees cavities which were then covered with a piece of wood. These cavities were twice the size preferred by bees, and the entrance was located one third below the ceiling of the cavity. The colonies expanded their brood nest at most down to two thirds of the way from the top and stored their honey for the winter in the top and bottom thirds. Honey stores in bottom third could easily be harvested without endangering the survival of the colony over winter. In those days a 100 per cent removal of honey stores, as often happens in conventional beekeeping, was impossible – and not even considered since there was no sugar available as a substitute.

Zeidlerei has survived until today in the southern Ural region and spread via Poland to Central and Western Europe. There are a number of zeidler initiatives that work on reviving the old tradition.

A logical next step in the development of beekeeping was to install tree cavities at ground level in order to avoid the strenuous tree climbing. This was the era of log hives, which was soon followed by the use of skeps in a great variety of shapes and sizes. All these hiving systems are characterized by fixed combs, i.e. the bees attach their combs directly to the walls of the cavity. From there they can only be removed by cutting them out. In all these instances a new colony was, at the beginning, a 'naked' group of bees, i.e. worker bees with a queen, which constructed their combs anew.

Only from the mid 19th century onwards inventions were made that characterize modern beekeeping up to the present age, and which have seen little developments since: moveable frames and consequently moveable combs, foundation that eliminates natural comb construction and artificial queen breeding.

There is no doubt that modern hiving systems fulfill the requirements of wild honeybee colonies even less than do the bee trees of the zeidlers. Hives are usually placed just 30 to 50 cm above the ground. Their volume in most systems is 100 litres or more, the entrance holes are rectangular (because they are easier to manufacture than round ones), and the orientation of the entrance depends on the location of the apiary, sometimes even facing North. Furthermore, they are not placed hundreds of meters apart but often side by side.

❧ Colony living in a tree at Mellifera

At the research apiary of Mellifera Association, nine living trees in the surroundings have been populated with bee colonies which are closely monitored and documented. Of crucial interest is the question, to what extent the feral colonies living high up in trees differ from those living in modern hives on the ground (controls). Analysing the pollen diversity in the honey reveals which plants they forage, providing information about the supply of the stores that allow them to survive the winter. Initial results have shown the water content of honey from the colonies living in trees to be 13.5 per cent, significantly lower than that of control colonies located on the ground nearby. However, it is already clear that irrespective of the large distances between the bee trees, forests cannot always secure a sufficient supply of nectar. Furthermore, the varroa burden can reach significant levels, since treating fixed comb feral colonies with organic acids is challenging or, in case of formic acid, even impossible.

〰 The location of the brood nest

There is one rule observed in all bee colonies, irrespective of the hive type they are living in.: whether be it a tree or a skep with fixed comb, or the movable frames of trough or Dadant hive systems, there exists one invariant constant: irrespective of the shape, the size, the type of colony expansion, bees will always locate the brood nest near to the entrance hole of the hive, whereas the honey stores are kept as far away from the entrance hole as possible. There are different opinions on why they arrange things this way: do they want to protect the precious food stores from predators, or is it easier to manage the circulation of air in the brood nest area when this is closer to the entrance opening? The answer to these question is irrelevant for the rule.

An maintenance opening on the side of a modern man made log hive allows a good view on the comb construction inside.

It is a matter of personal choice to describe these facts simply as 'concession to the requirements of the beekeeper', or even strongly criticize them to the point of total rejection. In our opinion we accept, as we do with all other farm animals and pets that it is no longer possible to fulfill some of the original requirements of animals if an intimate relationship between humans and animals is required. But at the same time this renunciation can be offset with a respectful attitude and regard for the integrity or even dignity of the animals. In bee-friendly beekeeping, this is achieved by allowing the colonies to follow their basic behavioural patterns, i.e. swarming and building their own combs.

Various hive types with moveable frame systems

Beekeepers love to spend hours discussing hive types. Each person is convinced they have found the perfect one. But in reality there are only three basic types: frame hives that are extended vertically by adding supers, such as the German standard Zander or Dadant magazine hives, and the Swiss hive. The latter is also managed by rear access, i.e. the frames are removed from behind, not from above like drawers in a cupboard. The Swiss beehive is ideal for keeping bees in bee houses. Unlike with magazine hives in the open, bee houses provide better shelter for the colonies from wind, weather, cold, and snow.

Then there are the vertical hive types, where extra space is added at the bottom, e.g. the Karnitz-Rings or the Warré Hive.

And finally there are the so called horizontal hives, where colonies are not expanded by adding space above or below, but to the right and/or to the left. This type includes the Top Bar Hive, as well as the many trough or long hives that can be found in Europe and Russia. Fedor Lazutin, a Russian bee researcher, lists eight different historical variations, which were developed independently from each other in the area of the former Soviet Union. There is a Warsaw Hive from Poland in the apiary at Fischermühle. Under the umbrella of the Mellifera Association, two horizontal hive types were developed: the Golden Hive and the Bee Box.

The Warsaw Hive is a traditional trough hive from Poland.

45

The Bee Box of our beekeeping colleague Heinz Risse from the Berlin regional Mellifera group is hung up in a tree in park in the city at an elevation of six meters.

On the one hand it is sobering to realize that what was thought to be a newly developed hive had actually already been built long ago, and in some instances almost with exactly the same dimensions. On the other hand it is satisfying to know that other experienced beekeeping colleagues also found the solution which we came up with is suitable and appropriate for bees.

The Bee Box arguably is the best documented bee hive as far as the internet is concerned. Its inventor, Erhard Maria Klein, has written two introductions to it which are well worth reading (see bibliography and web-address at the end of this book).

One thing is certain though: The diversity of hive types, and the innumerable discussions regarding their advantages and disadvantages, are proof for the fact that bee colonies take no or little notice of the shape and geometry of their homes. They display a formidable capability to adjust to cavities of the right volume in all kinds of possible and 'impossible' shapes.

The development of the Mellifera Golden Hive

Thomas Radetzki, the founder and long serving master beekeeper of the Mellifera Association, worked for many years with the Swabian Trough Hive, which is constructed like a half-timbered building covered with straw and plaster made from clay, cow dung, and ashes. This long hive contains vertically long frames and allows for the addition of small honey supers on top. This makes it possible to harvest varietal honeys. Besides working with this trough hive, he began to develop a second long hive with big frames. When he learned at a bee conference in the USA that the small hive beetle – a deadly pest for bee colonies – was drastically increasing in numbers, he started two initiatives. The first one was a political campaign which succeeded in banning imports bees into the EU from non EU countries where small hive beetles occur. The second was the development of a hive type which permitted the inspection without disturbance of the brood nest, which is where small hive beetles usually live. The reason for developing the new hiving system is because the beetle increases its rate of reproduction when disturbed. This led to the birth of the Golden Hive.

At the beginning of 2000, Radetzki introduced the concept for the new hive to some colleagues. Over a weekend the idea took on concrete shape, a blueprint was drawn and thanks to a beekeeping friend who also ran a carpentry shop the first 100 Golden Hives, with at the time the entrance hole in the centre of the box, were produced. When expanding the brood nest, frames are added in a symmetrical fashion on both sides in this big space, and the same holds true for the honey harvest. This makes it possible for the colony to grow undisturbed from its centre, without having to change the positions of combs. But of primary importance is that the whole colony remains together in one place, without being separated into different boxes/spaces. In particular, there is no queen excluder between the breeding area and the honey storage.

Of notice is the vertical dimension of the frame which was intentionally designed to match the size of a Dadant frame to make extraction of honey possible with any of the commercially available extractors. In contrast to the Dadant hive in which the long dimension of each comb is horizontal, in the Golden Hive it is vertical. Rotating the frame through 90° has far reaching consequences. Apparently bees match the diameter of the brood nest to the width of the available comb. This means that when using horizontally orientated frames, when the brood area is growing it can only maintain the rounded shape by including the first of the honey supers. That's why most beekeepers working with Dadant or other similar systems use queen excluders.

In a vertically oriented frame, the diameter of the brood nest is likewise adjusted to the available width of the comb. The result is that during brood nest development from spring to summer a perfectly round brood nest can not only be maintained on one frame, but also there is even room available for honey stores above the brood area (see the figure). If each comb contains some 500 g of honey on top of the brood nest and the brood nest is spread over 10 frames,, the colony will be left with 5 kg of honey after the honey harvest for the colony to supply itself.

On the other hand, in a Dadant system with a queen excluder, the brood nest is expanded to the extent that there is no space left for stores in the brood area. Therefore, if there is no more nectar available, it is absolutely necessary to feed immediately after the honey harvest. In a worst case scenario, when the harvest is followed by two to three weeks of rainy weather, thus preventing

❧ Without a queen excluder

According to biodynamic standards for beekeeping, queen excluders are prohibited and in the Golden Hive they are unnecessary. The brood nest will expand to a maximum of eleven to thirteen combs, honey is stored furthest from the entrance hole. The queen can move unhindered at will which contributes to the health of the colony. If required, the expansion of the brood nest can be limited by placing a comb full of honey next to it, since the queen will not 'jump over' this comb when laying eggs.

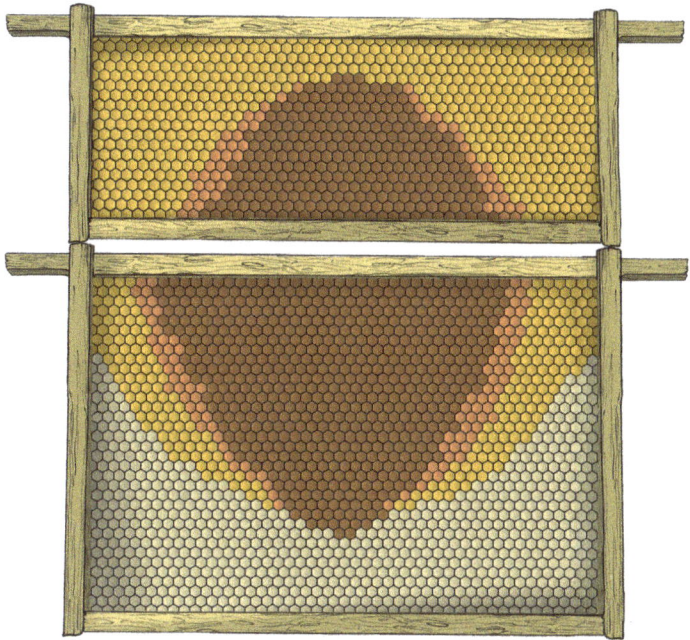

The bee colonies will establish a round brood nest in spring (brown area), which in dimension is adapted to the with of the comb. This is the reason why on a horizontally aligned Dadant frame (see illustration on top) the breeding area extends into the honey super frame.

Pollen (orange) is stored next to the brood nest, then honey (yellow). On the vertically aligned comb of the Golden Hive (see illustration on bottom half of the page) the round brood nest, the wreath of pollen, and the honey stores can be deposited on one and the same comb.

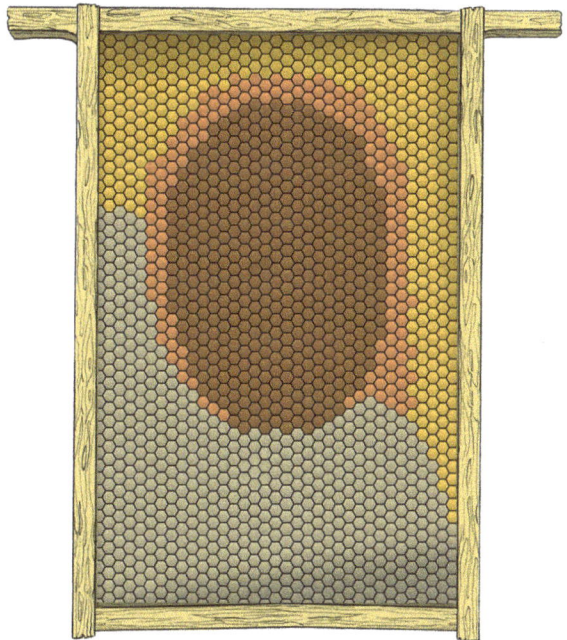

foraging, not feeding can lead to a colony's death by starvation.

When it comes to harvesting from the Golden Hive, the frames containing only covered honey cells, no brood, are removed and put into a comb transporting box. And at this stage another advantage becomes obvious: no lifting of heavy wooden boxes is involved (in the case of Dadant up to 15 kg or more). A single frame full of honey from a Golden Hive weighs 3.5 kg when fully capped and frames are removed during inspection one at a time. Not least for this reason, the Golden Hive has been nicknamed by some as the 'Lady's Hive'.

The honey store over the brood nest is ideal for the development of the colony, as it ensures a continuous supply without the stress of running short. The dimensions chosen for the comb are of an optimum size for natural comb building, and they enable a large, uninterrupted brood nest area. And in contrast to all other systems with large frames for the brood box, which requires in most cases a second set of differently sized frames in the supers, the Golden Hive manages with only one frame size.

Another special characteristic of the Golden Hive is that all proportions are set in compliance with what is known as the 'golden section' or 'golden ratio'. This is true both for the dimensions for the comb/frames, as well as for the width, height, and length of the hive box. The golden section is a universal design principle of living nature, which can be observed in plants, animals, and the human body. It describes a balanced ratio of divisions in the relationship of two dimensions with each other, which we experience as perfect harmony.

Further developments of the Mellifera Golden Hive

A symmetrical expansion of the bee colony and the filling of honey combs on both sides of the entrance hole are both aesthetic as well as bee-appropriate. But one drawback of this method is that in the summer a long, cigar-shaped brood nests can frequently develop and stretch over up to 16 combs, making honey harvest from these frames impossible. Therefore, in the second production run of Golden Hives the entrance hole was positioned towards the end of a long side of the hive. This break of the symmetry made the harvesting of honey from the part of the hive furthest away from the entrance easier and more productive compared with hives with entrances at the centre of the long side of the hive. Since some beekeepers did not like the beak in symmetry, they switched the entrance hole to the small side of the hive – a management method that works perfectly well, but results in a slightly stronger tendency to swarm.

The current version of the Golden Hive is characterized by a few novelties. The most important of these are:

▷ a closed floor, which makes it easier for the bees to regulate the microclimate within the hive

▷ a sophisticated method for the aeration of the hive which largely prevents condensation in areas where bees cannot regulate the microclimate

▷ three semi-circular entrance holes, which can be opened or closed as required.

The Golden Hive has been designed for extensive honey production, and is in compliance with the biodynamic standard. The annual expenditure of time for the management of a colony is reduced, since the colony is spread over larger, but fewer combs compared to other hive systems. If an inspection of the brood combs is necessary, it is possible to lift those immediately, without first having to put aside honey supers. This reduced need for inspections, and their shorter duration, means less stress for the colony, which is mirrored in the docile temper of the bees when the hives are opened. Until today, some 3,000 Golden Hives are in use, and their number is growing.

Construction and components of the Golden Hive

The new version of the Golden Hive is designed to hold 22 frames in which the colony can develop its brood nest and keep its stores. For winter or when establishing young colonies, it is possible to restrict this space to eight or ten frames with a division board made from wood or straw. A waxed cloth and an insulating top board protect the colony against heat in summer, as well as cold in winter. A metal cover lid protects against rain. The high floor allows for two trays with a bee-proofed grid needed for diagnosis of the mite load.

The three entrance holes satisfy various requirements: The central one

permits traditionalist beekeepers to manage colonies the original way. The entrance holes near the corners allow for easier colony management for beginner beekeepers or, alternatively, allow for raising two young colonies in their first year in the same hive. For this, of course, it is necessary to insert a bee-proof dividing board which completely separates the box into two compartments (see page 56 for more).

The top-bars of the frames are shaped with an edge to facilitate the direction of the comb construction. Inserting a starter foundation strip is therefore unnecessary. However, also

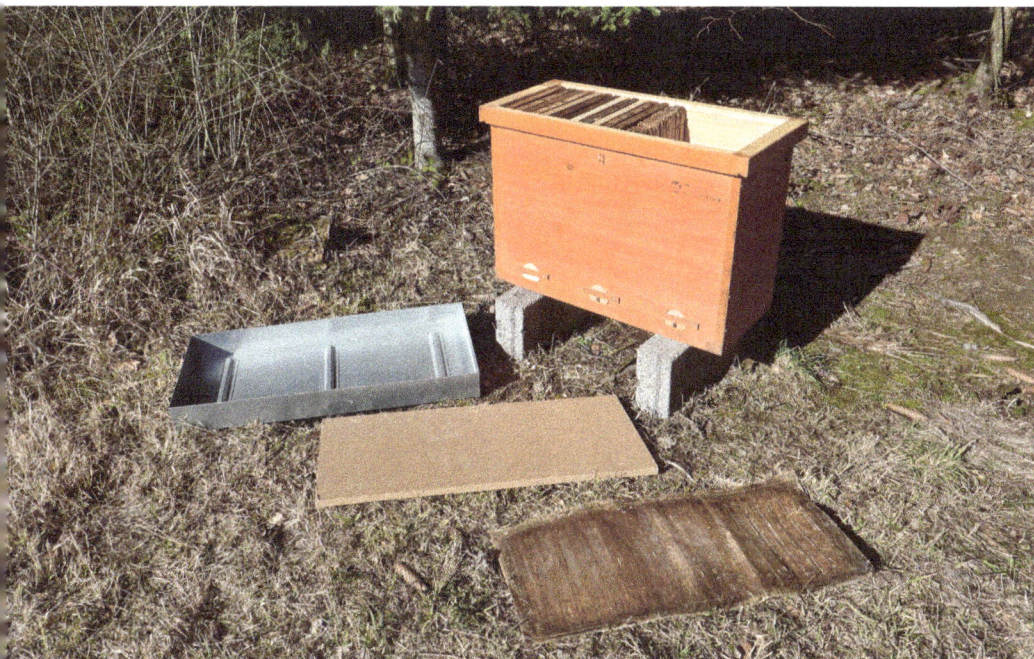

The metal sheet slip lid keeps off rain, the soft fibre board provides thermic insulation, and the waxed cloth is placed as a protective cover onto the top of the frames.

commonly in use to provide direction for the comb construction are top bars with a notch, into which a thin starter strip can be inserted and which can be melted in the first framing wire to enhance its stability.

'Bee Space' is defined by the average distance between two combs when these have been built naturally. The bees make use of this gap, which is 8 (+/-2) millimetres wide, to walk on both opposing combs, without getting in each other's way. They also keep this space free of comb between the frames and the wall of the hive. Spacers comprising mushroom shaped nails, or mushroom shaped wooden pegs, are placed onto the frames for this reason (see illustration on page 58). They ensure the correct spacing of the frames after they have been pulled out for inspection and then replaced. Four to five lines of stainless steel wire provide stability for combs and foundation. They prevent the tearing of soft, fresh comb when a lot of honey is stored in them.

The problem of condensation which leads to the development of mould in spaces, which are not covered and ventilated by the bees, is common for many types of hives. In the new version of the Golden Hive this is mostly prevented by means of three ventilation holes (covered with bee proof wire mesh) in the bottom board, and a thin ventilation slit on top of the narrow side. Both openings can, with very few exceptions, remain open all year long. Only when treating for varroa should they be closed (see page 119 on this). To achieve this, the holes in the bottom can be simply closed by putting a piece of plywood or cardboard on top of them.

When feeding from a bucket placed in the empty part of the hive, the top ventilation slit should also be closed by means of the provided cover. The honey in the bucket, or other substances with a typical scent like herbal teas, will attract bees from neighbouring hives. To find the entrance hole is an easy task and robbery is particularly likely to occur if neighbouring colonies are being fed with the same food substance at the same time. As is shown later, in biodynamic apiculture the sugar solution is enriched with herbal teas and honey. This solution is much more attractive than plain white sugar and water.

If a colony is large enough to need all 22 combs, i.e. if there is no space in the hive not used by the bees, then the ventilation holes in the floor board and the slit in the top of the wall of the hive top should be closed in order to prevent draughts and the emission of sweet scent from the hive that might invite robber bees to check for the availability of honey.

๛ Influencing the direction of the building of comb

Bees never build combs next to each other in a random manner. Instead function and use are always considered. The top consideration is the management of the microclimate by means of the space between the combs. This does not mean that the combs will always be built exactly parallel. But the parallel construction is essential, if the beekeeper wants to lift combs for inspection and honey harvest. Therefore comb guidance with a ridge or a starter strip, or an edge on the top bar of the frames is necessary.

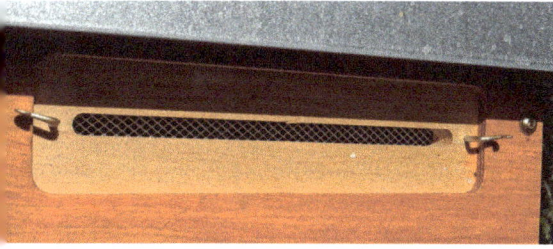

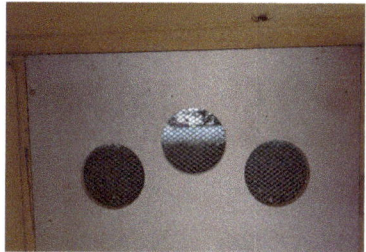

A slit for aeration (top and bottom left) can be opened or closed according to necessities. Three holes in the floor (top and bottom right) allow for a good circulation.

The hive is covered and closed by means of a piece of wax cloth made from linen or cotton. These natural materials are bee-friendly, and they make inspection easier. As an alternative, it is possible to use a tightly woven piece of cotton or linen, which the bees will then cover with propolis themselves (see page 154 for this). If the reason for opening the hive is simply to expand the colony or add frames for honey, then the cloth is folded back only to the extent required for this addition. All the other frames remain covered. As a result the disturbance of the bees is minimised.. This is important especially for maintaining the optimal temperature and microclimate.

In order to limit the space occupied by bees, two types of division boards are available: two dividers made from insulating fibre board reinforced with a layer of solid wood are supplied with purchased Golden Hives. When hanging next to each other, but displaced relative to each other, they provide a complete seal between the colony and the empty space not occupied by frames. The second excellent option is a division board made from straw purchased separately, and which also provides a tight seal. Straw can absorb and release moisture. This is important for the climate within the hive. It takes up water in early spring and may release it during the hot summer days. Both types of boards allow colonies for optimal regulation of air circulation in the spaces between combs.

A partition board made from straw or a soft fibre board are important to regulate the temperature. It can be used to limit the space for the colony in an optimal manner, when the hive is not completely filled with comb. It partitions off empty space of the Golden Hive, which may not be in use during the winter season.

❧ Measurements and parts of the Golden Hive

▶ The Golden Hives supplied by Mellifera are made from glued wood boards of Weymouth pines. Linseed oil varnish is included for application by the buyer.

▶ The hive has an internal dimension of 790 x 302 x 540 mm (L x W x H), and outer measurement of 840 x 350 x 580. The weight empty is 14 kg.

▶ By adding or removing frames it is possible to enlarge or reduce the space for the colony. The hive can be filled with a maximum of 22 frames. The shipment includes 21 frames including 7 mm nickel plated mushroom shaped nails to create the bee space for the buyer to fit. Recent experience has shown that the addition of a 22nd frame is hard to achieve, since bees are adding tiny amounts of propolis to the nails. Therefore, the shipment includes 21 frames, only.

▶ The box has three round entrance holes, ventilation holes in the otherwise closed bottom board, and a ventilation slit at the top of one side.

▶ These additional parts are included with the Golden Hive:

▶ A wax cloth (400 x 900 mm), to cover the colony, made from certified organic cotton and organic beeswax. It is placed directly on the top of the frames.

▶ A cover board made from 18 mm particle board, which is placed on the wax cloth for insulation purposes.

▶ Two dummy boards for limiting or enlarging the space inside the hive; they are a sandwich construction of 20 mm particle board and 2 mm poplar plywood.

▶ A metal roof, without rivets, with integrated spacers and a UV stabilised elastic cord.

Feedback from many beekeepers using the Mellifera Golden Hive result in occasional minor modifications and improvements. What remains constant, however, is the internal dimension and the size and total number of frames.

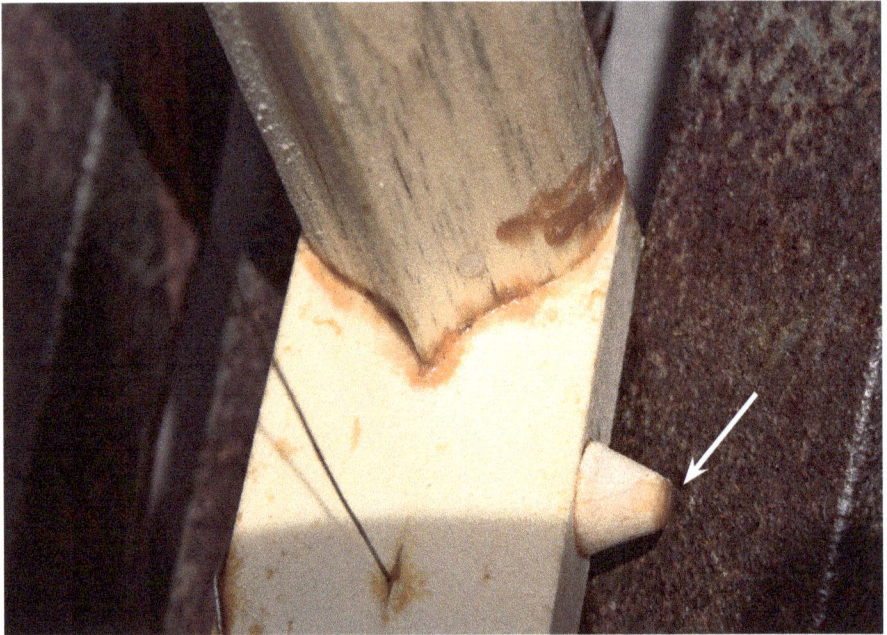

The keel shaped profile of the top bar of the frame encourages the bees to build the comb in a perfectly aligned manner into the frames. Mushroom heads (see arrow) ensure the maintenance of the so called bee space of 8 to 10 mm between the combs. This space allows the bees to move over the comb without getting into each others' way.

Useful accessories

We recommend that in addition to the Golden Hive it makes sense to purchase a box for transporting combs, and a swarm collection box. The comb transport box is needed to take honey combs from the apiary to the extraction room and to store them after extraction. Both boxes have room for eight combs. Thus, it is possible to keep a swarm or a split/nucleus with a queen (see page 92) in one or other of the boxes over the summer and to let a colony develop. However, the thin, light, plywood walls make these boxes unsuitable for overwintering such colonies, as they would have difficulty maintaining the required temperature.

Plywood and a metal wire mesh ensure lightness and ventilation of the swarm collection box. A funnel can be attached if an artificial swarm is to be formed (see picture below). The bees are brushed from the frames into the funnel. They will walk with their queen immediately through the entrance hole into the darkness of the box. At the strap of the lid (see picture above) the bees will form a swarm grape quickly.

The comb transport box (picture to the right) has room for eight frames. It is made from light weight plywood and thus easy to carry to the honey extractor. With the entrance hole (at the bottom) the box allows for temporarily housing a young colony in summer.

The original and copies

Since the number of people who keep bees in Golden Hives is growing, other vendors decided to produce and sell it. Right from the beginning, Mellifera Association intentionally never tried to protect or register the design, which is why copying is possible without consent. However, some of those hive builders felt it necessary to ask for permission produce the hive. Our responses varied depending on the inquiry. If requested for strictly personal use, we always agreed unconditionally. If the intention of reproduction was to earn money, Mellifera asked for a share of the profits to be transferred to its charitable association. There are manufacturers who comply with this request. And there are others who offer the hive on a commercial basis without any contact to the Mellifera Association.

You should closely compare the size of the frames, as some suppliers use different measurements. Thus, frames cannot simply be exchanged between differently sourced boxes.

Mellifera's website lists a number of suppliers of Golden Hives, as well as offering a manual for those who want to construct one themselves. This includes cutting guides for wood from timber merchants or DIY outlets (see page 168).

On an almost daily basis, beekeepers call us or send emails asking for advice on managing their hives, irrespective of where they have obtained their hive from. The requests are understandable, since the Golden Hive was, after all, developed in our experimental apiary, and at times we managed more than 100 of them. We believe that we have the most experience that is available. Nowadays, the numbers of inquiries exceed our capacity to respond, not least because other manufacturers share their own ideas of how to manage a hive more efficiently or economically – yet their customers come to us for advice.

The primary focus of this book is the description of the original hive from the Mellifera Association. Our hives are manufactured in a social therapeutic workshop in Germany, and they are branded exclusively with the Mellifera logo.

All hive systems are considered to be Mellifera Golden Hives, if the outer measurement of the frames is 458 mm tall, 285 mm wide, with 21 to 22 of them fitting into the hive. The shape of the top bar of the frame, the clearance in the bottom area, a closed as opposed to a mesh floor, the location of the service flap for cleaning the floor, or the thickness of the wooden walls are of some relevance, but they are of subordinate importance for the basic style of management. Because of this we are confident that this book will be of help to all beekeepers who are housing their bees in a Golden Hive.

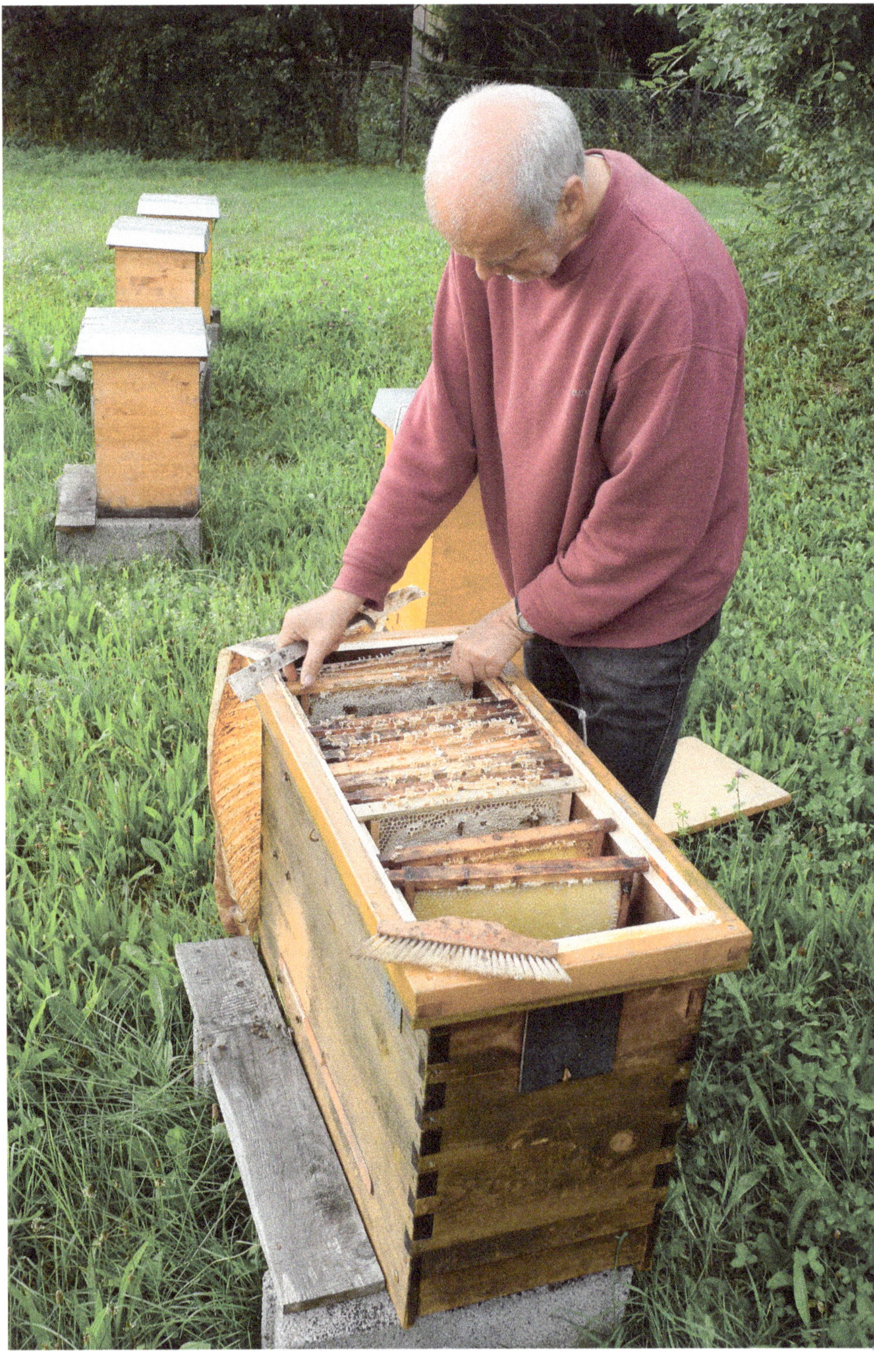

Beekeeping: How to begin

Before going out and buying one or more Golden Hives, a number of issues need to clarified. Otherwise frustration and irritation will be guaranteed.

The first point is the issue of population density. This is not easy to answer, as it may not be possible to find out reliably how many bee colonies are already located within a radius of three to five kilometres – the flight distance of a colony. If the number is high (above 20 colonies per km2) one may even have to consider not putting any more colonies into that area. Particularly in urban areas there are nowadays so many colonies, that there is no longer sufficient pollen, and in particular nectar, available for the bees to adequately feed themselves. Obviously it is possible to help out colonies during gaps in the honey flow by supplementary feeding of sugar water. But even if this is improved and made tasty by the addition of honey and herbal teas, it will never be a complete substitute for nectar and honey.

We recommend that like-minded people get together and form a bee-keeping interest group. Often it is more fun to collectively work with the colonies, to discuss various angles and opinions on measures to be taken, and to take decisions collectively. This also take away tension during the swarming period. Many will be relieved if they have only to check every two to three weeks. Because otherwise there are no free weekends from early May to end of June. By contrast, if I keep bees together with two colleagues, I only need to check for signs of swarming every three weeks.

Beginners are strongly advised to attending a course on bee-friendly bee-keeping. One will often search in vain for help among traditional beekeeping associations on how to manage the swarm impulse, because they will teach only methods on how to suppress this elementary character trait. Similarly, finding advice on natural comb construction will be difficult, since conventional beekeepers only work with embossed foundation. And the mating of young queens will be viewed with scepticism, since the majority opinion will hold that it is not possible to successfully keep bees without artificially bred queens. And last but not least, if one keeps bees in a Golden Hive instead of a standard frame hive, it is tiring having to deal with critical comments at ordinary beekeeper association meetings.

On the website of Mellifera Association (address see page 168) you can find various training courses on offer, which will deal with various management aspects of beekeeping in Golden Hives, but also in a number of other types of hives. These courses bring together interested people who are like minded on questions of dealing with the swarming urge, natural comb construction, and the avoidance of artificial queen breeding. Regional Mellifera groups at many locations enable a constructive and productive exchange among friends who share the same aims and objectives in beekeeping.

You should also consider the amount of time that should be devoted to beekeeping. A rough estimate is that you will need at the beginning of your beekeeping career approx. 30 minutes per hive per week. That does not sound like a lot to begin with, but tasks that need doing are not spread evenly over every week and every colony. Instead there will be phases, when a lot needs to be done, and other more quiet periods. The amount of work is dictated by the requirements and needs of the colonies. For example, , if the loss of swarms is to be avoided, or if one is planning an artificial swarm (see page 88), during the swarm period weekly inspections are necessary. Production colonies will require at the beginning of certain nectar flows a speedy expansion, which also requires work input. And if there is honey to be harvested, it should be centrifuged and bottled with as little delay as possible in order to prevent quality loss.

Frequently it is necessary to feed swarms and young colonies regularly, so that the comb can be constructed quickly. From July onwards it is necessary to monitor the varroa count on a monthly basis, and it may be necessary to treat against varroa. And from August onwards it will be necessary in most instances to provide supplementary feeding to help the colonies put in sufficient stores for the winter. These tasks are all fairly easy to manage, but they should also fit into the schedule of other demands on the beekeeper concerned.

Wherever hives are located it will be necessary to gain permission from the owner of the land in question. In addition the regional veterinary authority[3]

3 In the UK beehives should be registered with Fera's national bee unit (http://www. nationalbeeunit.com

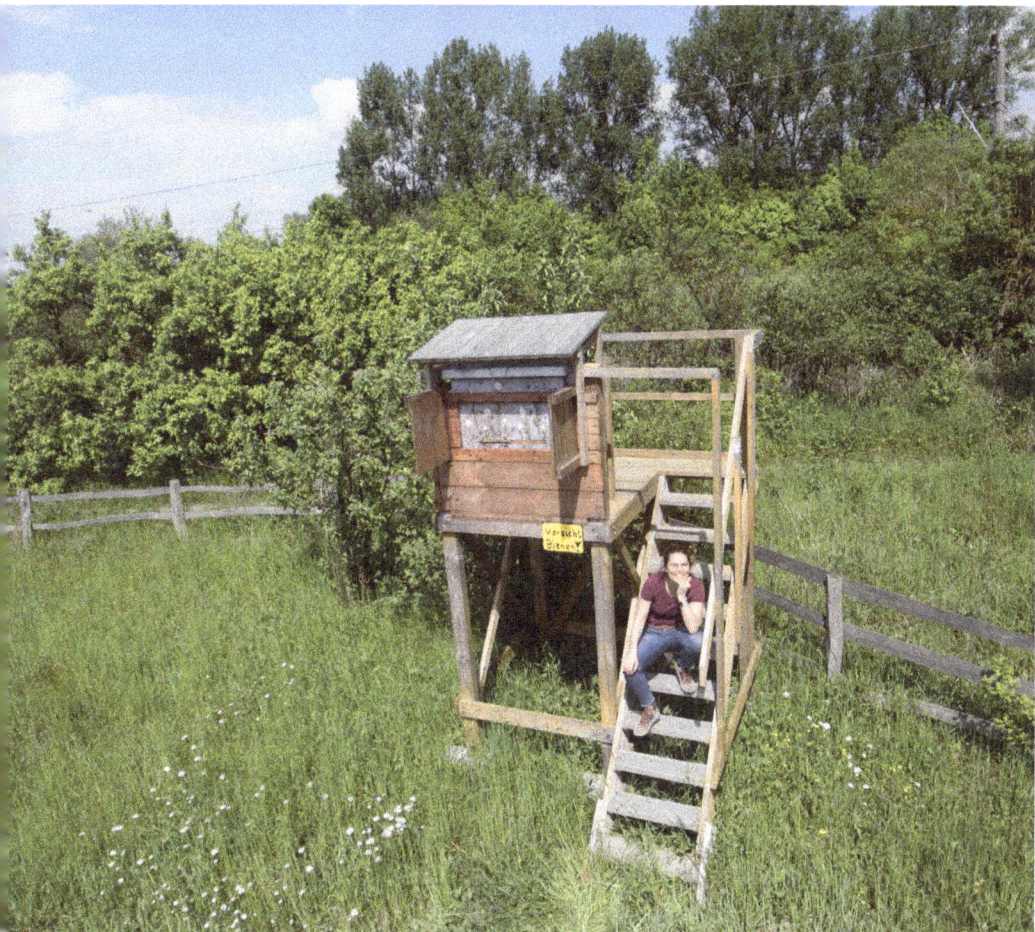

Chef and hobby beekeeper Sarah Wiener has constructed raised stands for a number of Golden Hives at selected locations. They are spaced well apart from each other and placed in surroundings that provide plenty of pollen and nectar and thus promise a good colony development.

must be informed about all relevant beekeeping activities[4], so that speedy action is possible should there be a disease outbreak. Registering one's hives with the authorities includes the obligation to treat against the varroa mite.

Informing land owners and the neighbours, too, should not be seen as a bothersome duty, but rather as an opportunity to inform others about the situation bees are in. One may even invite them to come and visit the colonies – it is not an infrequent occurrence that bees manage to charm the sceptics.

4 In Germany this is a legal requirement. In the UK it is highly recommended

Preparation and siting of the Golden Hive

The Golden Hive is a carefully designed and skilfully crafted home for bees. Therefore it is worth protecting it against rain and weather by treating it with a bee and environmentally friendly finish. Two paint coats should be given initially. The moveable parts such as the covers of entrance hole, the ventilation slit and the maintenance hatch, as well as their rims and slots should also be painted.

A paint coat with pigments will provide the required UV protection. There is not only no need to paint the inside of the hive but also it should not be done. Bees will over time cover the walls, the frames and combs with a layer of propolis (more on this on page 154). The amount of propolis applied will vary from colony to colony. But one may rely on the fact that the bees certainly know their needs regarding the hive's internal microclimate.

If you already keep bees, you may rub the inside of a newly made Golden Hive with propolis dissolved in 90 percent alcohol, a drop of lemon grass oil or a bunch of Melissa. This will help the bees more quickly to feel at home in their new lodgings.

A second application of paint will become necessary after four years at the earliest provided that high quality natural paint has been used initially. If water-based or boiled linseed oil varnishes are used, they may be applied in the open, even when there is a colony in the hive. Above all, whenever a hive is empty it should be thoroughly checked, repaired, and, if necessary, given a new coat of paint. As bees have a very sensitive sense of smell, a new paint coat should be allowed to thoroughly dry before a colony is hived in the box.

The working height can be adjusted to the height of the beekeeper in order to make hive management easier and thereby limit back strain. The height is optimally adjusted when it is possible to look into the spaces between the combs without having go bend forwards excessively. At the same time it should be possible to pull out a frame without having to stand on tip toes. Usually Golden Hives are placed on concrete blocks or wooden racks with a height of between 25 and 40 centimetres. It is extremely important to ensure that the hive is positioned perfectly vertically, otherwise the bees might start comb construction at the top of one frame and end up at the bottom of the neighbouring frame.

The coating material for the cover of the outside of the Golden Hive must contain sufficient pigments to prevent weathering. But apart from this there are no limits for creativity.

Maintaining a distance of at least 40 metres between individual colonies will significantly reduce drifting of foragers. We especially recommend this minimum distance to hobby beekeepers with few colonies and sufficient space within their own gardens or at a nearby site.

Beekeepers are also inclined to philosophise for hours about the correct, but often we are obliged to adapt to reality, particularly when colonies are placed in our own garden with neighbours close by. In order to avoid cold winds blowing into the hive in the cold months of the year, helps if the hive entrance does not point North. Similarly West orientation is not ideal, because this is usually the weather side from which comes driving rain. A semi-shaded site is ideal, but bees can cope with sites in full sunshine, provided they have an

opportunity to collect water nearby (e.g. a few 100 metres). If the entrance hole faces South East, the bees will be woken up in the morning by the first rays of sun.

Another important aspect is related to neighbours. The Golden Hive should be situated in such a way, that the bees will already be high in the air when they cross over your neighbour's land. This is quite important for amicable coexistence, particularly in densely populated residential areas. A lot can be done in such areas to avoid dreaded bee encounters. Bushes and shrubs next to the hives can act as barriers for flying bees. Should there be a footpath or a street in close proximity to the entrance holes, one can place a two meter high fence or hedge which will ensure that the bees to take off at a steep angle and return to the hive following the same path. They will handle such obstacles without any difficulties. We have experienced that due to such arrangements none of the neighbours actually noticed of our little friends.

Whether a site is good or bad for the bees depends to a large degree on the availability of flowering plants. Thomas Seeley has demonstrated that colonies from his apiary obtain nearly 50 per cent of their honey within a flight radius of 1.6 km. But for the remaining 50 per cent they may fly as a far as five(!) kilometres. The flight radius for foraging is on average 2.2 km. Hence a site should have sufficient forage within a radius of three km, preferably excluding any industrial agriculture whose pesticides could damage the colonies.

One also has to keep in mind that only a few plant species permit the collection of large quantities of nectar. Usually good sources include cherry, dandelion, sycamore, apple, false acacia, oilseed rape, lime, horse chestnut, as well as honeydew from various deciduous trees and spruce and silver fir. But the changing seasons of the year and the variation in precipitation can have strong impacts on the quantities of nectar produced by plants.

The purely subjective aspect is also very importance in choosing a hive site. Beekeepers should be able to feel comfortable in their apiary. If from whatever source or for whatever reason they experience stress , they will certainly pass this on to the colonies. But if they are able to work in a relaxed manner, the bees too will behave in a friendly way. On the other hand, one should also take into consideration that factors that irritate humans have no relevance for bees and vice versa. For example, traffic noise bothers the beekeepers but not the bees. Small vibrations of the ground are hardly noticeable to humans, but bees definitely detect them. It makes sense to gather impressions of a proposed apiary site over the course of a full year. A site which may be inviting on a warm summer's day may feel bleak or even threatening on cool and grey days in the autumn. And although studies have shown that colonies react to their location based only on availability of forage – sparse or abundant, early or late flying opportunities – when it comes to docility and or proneness stinging the beekeeper is the decisive factor.

A swarm moves in

The best time to obtain a swarm is during the swarming season, i.e. between mid-April and end of June. The best options are a natural swarm or a pre-emptive artificial swarm from a beekeeper in the neighbourhood. We recommend starting with a natural or a pre-emptive swarm, because in these cases all the bees originate from the same colony. Since most beekeepers nowadays multiply their colonies by splitting them into the their own hive systems, in some locations it can be quite difficult to find a split that fits a Golden Hive. Beekeepers refer to colonies that are without frames or combs as being 'naked'. So you should look for a swarm, an artificial swarm or a naked swarm. It is advisable to make early contact with the beekeepers and to let them know you are looking for one or more swarms. In many instances the chair of the local beekeepers' association may help, as he or she would most likely know who might have swarms on offer. Since most beekeepers in a region know each other, if you draw a blank with your first contact, ask other possible contacts for swarms.

Hiving a swarm in a prepared Golden Hive should ideally be done in good weather, with outside temperatures suitable for normal bee flight, otherwise the bees may congregate at the bottom of the hive box instead of ascending to the top of the frames.

∾ The swarm moves in

In order to house a swarm in the Golden Hive, it needs first to be prepared. If it has never housed bees before, it should be intensively rubbed with a bunch of Melissa, or sprinkled with lemon grass oil, or painted with propolis tincture (see page 155). Bees do not like freshly finished wood. The walls of already used hives are carefully scraped with the hive tool and scorched with a blow torch.

Following this, and depending on the size of the swarm, five to seven empty frames are placed in the hive above the future entrance hole. Beginners are advised to select the hole furthest from the ventilation slit. A little honey is smeared onto the tops of the frames to later attract the bees to quickly ascend, following which they will assemble similar to the cluster formed in the swarm collection box.

A follower board is put next to the outermost frame, and a wax cloth folded to the required size is used to cover the frames. The remainder of the box remains open. The swarm is shaken from the lid of the swarm collection box into the open space. The bees will then hurry towards the dark part where the frames are.

However, it is much more enjoyable to allow the swarm to run in through the entrance hole, i.e. not via the roof, but via the main door. The Golden Hive is prepared in the same way as outlined above, the frames are put in, the wax cloth is folded back to cover only the area with frames. A board is placed at a slightly downward sloping angle in front of the entrance hole. The swarm is shaken onto this board from the swarm collection box.

Once the bees, barring a few stragglers, have moved into the dark area with the frames, the wax cloth is unfolded to cover the empty area of the hive, too. A single frame, placed in the middle of the open area, will provide sufficient support for the cloth to prevent it from falling into the cavity. Following this, the composite particle board is placed on the cloth and on top of that the metal lid. They are then secured with the elastic cord.

There are two possibilities of putting a swarm into a hive. The classic method consists of shaking the bees into the box, the more impressive way is letting them run in by themselves (see page 71).

It is very important to check at the latest on the following day whether the colony has settled at the correct place in the Golden Hive. Should this not be the case, then the frames are shifted to the location where the bees are hanging in all likelihood from the wax cloth (without frames). Before moving the frames it is necessary to shake the bees down from the cloth. The alternative option, i.e. to bring the bees to the frames, rarely works once the they have chosen a different location within the hive. The next step is to place division boards on both sides of the frames in order to limit the available space for the colony. This will encourage the bees to find their way from the floor to the frames. As soon as the colony is nursing its first batch of brood, the frames can be shifted to whichever place in the hive that the beekeeper deems most appropriate. The bees will always stay on the brood.

It is possible to invite family, friends, and children to watch the magic and beautiful spectacle, when the bees are allowed to run through the entrance hole into the darkness of the hive, hesitating at first, but, once the queen has found her way into the box, in a hurried melee. There is no need to be afraid of stings when a new swarm is running in.

Feeding new colonies

In most instances colonies require initial feeding since the nectar flow may not be sufficient for a speedy construction of combs. Feeding can be started on the first, though preferably on the second evening after the hiving process. Prime swarms with old, mated queens construct combs very quickly. In less than three weeks they will have constructed combs in eight to twelve frames. This is an advantage because after two weeks colonies begin with the construction of drone cells.

A characteristic of a young colony with an old queen is that it will rarely be tempted to swarm again. Should the colony not be satisfied with its queen, it quietly supersedes the queen without any outward signs, i.e. replaces the old queen with a young one.

Feeding a real prime swarm is done by providing a single instalment of five litres of sugar syrup. Pre-emptive swarms (see page 88) receive an initial three litres, and the same amount again after five to ten days. If there is no nectar flow, supplementary feeding is required in both instances.

It is best to feed in the evening hours after the foraging flights have stopped, since this reduces the chance for robbing. It is always amazing to see that large swarms can take up all the feed within a single night. It is placed next to the dummy board in the vacant part of the hive. A bucket with a mesh lid for liquid feeding is placed onto two wooden strips. The bees then walk underneath the bucket and sip the food from the mesh with their tongues (probosces). Make sure that the lid is tightly closed, otherwise air will seep into the bucket and the food will leak through the mesh onto the hive floor. Alternatively any container or bucket with the top open can be placed inside the hive.

In the case of smaller colonies, it is advisable initially to limit the portions fed. A continuous supply of food, for example via an upturned feeding bucket with a mesh, or repeated smaller feeds, are best (see page 94). When later on larger areas of combs allow the absorption of large portions of feed, it is possible to give five, or if necessary even 10 litres, e.g. when feeding large colonies for winter.

∾ Preparation of the Food

We begin by preparing the herbal tea. Chamomile blossoms and a pinch of salt are essential. In addition we add all the medicinal herbs we can find in the garden.

Crystalized sugar, i.e. refined sugar from certified organic or biodynamic production is added to the tea. The choice of sugar has nothing to do with nutritional content. In that respect it does not differ from that of conventional agriculture. However, conventionally grown sugar beet is produced with pesticides and fungicides with their associated harmful impact on the environment.

Once the temperature of the tea-sugar mixture has fallen below 40°C (hand warm), 10 per cent by weight of honey in relation to the quantity of sugar added is stirred in. The honey should come from your own production or from a safe source. Temperatures higher than 40°C destroy the valuable enzymes. We cannot recommend using honeys from supermarkets, as these are often contaminated with American foulbrood spores.

The composition of the food varies according to the seasons of the year:

▶ up to the end of June: 1 part herbal tea, 1 part sugar, 10 per cent (by weight) honey

▶ from July: 2 parts herbal tea, 3 parts sugar, 10 per cent (by weight) honey. A tried and tested blend consists of 2 litres of herbal tea, 3 kg sugar, and 300 g honey.

If only a few bee colonies require feeding, the simplest method will be to prepare the food fresh whenever needed. The hot tea is added directly to the sugar. The process of dissolving it will reduce the temperature quickly, so that the honey can soon be blended into the mix.

To avoid bees drowning in the open feed, some straw, or, even better, hay still smelling fresh, or wood shaving from an organic source, are put into the open bucket. If the bucket is new and smooth, adding branches of spruce resting against the sides provide a welcome help for bees entering the bucket. Another option is to roughen its sides with sandpaper. The bucket should be placed against either a wall of the hive or the dummy board in order to make it easier for the bees to climb in and out. Containers with fold-over rims are impossible for bees to climb into.

Straw or hay are used as landing substrate for the bees in the feeding bucket.

In many locations bees do not find sufficient nectar in order to build up their winter stores, therefore feeding is necessary. With young colonies and young queens the winter stores are built up during the entire bee season until mid-September.

∾ Fondant for exceptional circumstances

Sometimes fondant is given as a feed and likewise presented in an open bucket. Bees will accept this emergency food only if neither nectar nor liquid feed is available. Fondant does not support the urge to build comb. That's why we recommend its use only in specific situations, e.g. when the beekeeper goes on vacation, or otherwise when it is impossible to visit the colonies on a weekly basis and feed as required.

Bee keeping through the year

In our part of the world bee colonies live in cavities or hives, which provide shelter against rain, snow, excessive heat and cold. Nevertheless, they are much more exposed to the weather compared to we humans living in houses with heating, air conditioning and electric light allowing us to extend our day length at will. But bees, too, have the capacity to heat, cool, ventilate, reduce moisture, clean and do many other things in their enclosures. And similar to our way of maintaining different temperatures in different rooms, bees do the same for the spaces between different groups of combs. In order to do so they must co-operate in groups, and great effort is required from each individual bee involved in the task. It is noteworthy that the biography of a worker bee is characterized by a definite succession of different tasks, But when it comes to heating the brood nest and the colony, all the bees are involved, irrespective of their age.

Honey stores are required in order to provide sufficient energy for fine tuning of the nest temperature, and not only in winter. Honey helps the bees to bridge gaps when, due to rain or low temperatures, no nectar can be collected. Martin Lindauer, one of the best known bee scientists, links this particular capacity of the bees, i.e. adjusting the hive temperature, to their successful evolutionary history over millions of years.

◀ The busy activities during summer and autumn are followed by a quiet period over winter. The bees sit in a tight cluster inside the Golden Hive. Holes in its bottom ensure sufficient aeration even in cases, where snow blocks the entrance hole.

End of winter and spring

Beekeepers differ in their definition of the start of the beekeeping year. There are reasons to link it with the breeding of healthy winter bees, which carry a bee colony through the cold season. For us, the beginning happens when towards the end of the winter the winter cluster is slowly breaking up and the first brood is being reared. At this time the winter bees are more and more replaced by summer bees.

On the first warm days in late winter we start to see them in the open flying in and out the hives, their humming serving as a harbinger of spring. The cosy compactness of the cluster, easily visible with the help of a thermal camera on cold days when not a single bee leaves the hive is now released. The winter bees share the various tasks inside the hive and outside in the surroundings.

In early spring, the flowering hazels, the first noteworthy pollen source), markedly increase brood production. Sudden cold spells are frequent during this period and provoke significant efforts to keep the still small brood nest constantly close to 36°C. When outside temperatures rise to 11°C or more, bees like to leave the hive in order to empty their rectums, collect pollen from the hazel bushes as fresh protein for the brood, and bring in water.

At the time of the flowering of the goat willow, the size of the brood nest is significantly expanded. In a few weeks the queen will reach her maximum laying capacity of 1500 to 2000 eggs per day. In 21 days these will develop into young bees. For the older winter bees this marks the beginning of the most strenuous part of their lives.

In these first weeks of spring, a relatively small number of winter bees is available to feed a growing quantity of brood. The German term 'Durchlenzung', the period between the dispersal of the winter cluster and the start of the main nectar flow, refers to a critical period in March when many worn out winter bees die. By now, a good supply of protein for the brood is essential to ensure the hatching of strong and healthy young bees to increasingly replace their old winter sisters.

But more than a varied supply of pollen is required in spring. In addition to the willows, other early flowering plants such as snowdrops, crocus, winter aconite, blackthorn, and Cornelian cherries provide fine nectar which stored in cells is unmistakable due to its moist glitter. As capped honey cells are opened by colonies only in an emergency, busy foraging bees, in a natural setting where no honey is removed from the colony, may already begin storing the first winter provisions for their sisters who will emerge in the autumn many generations later. In contrast, as beekeepers we often start thinking about winter feed for our colonies towards the end of the bee season.

Blossoms of orchards, in particular cherries, apples and dandelions provide the first significant nectar flows of the still early bee season. From this point onwards colonies have two options: either they develop the swarm urge, or they serve as production colonies that collect large quantities of nectar and provide us with the much appreciated spring honey.

🌱 Monitoring the flowering periods

If we care for the wellbeing of our bees, we should pay close attention to the weather as well as to the development of both flowers and bees, drawing the correct conclusions from these observations. For example, it makes sense to get to know well the most important nectar providing plants, and note the sequence in which they flower. While the calendar date on which plants go into bloom varies depending on weather and season, the sequence remains the same. Our colleague and friend Albert Muller is sure that armed with this knowledge it is possible quite precisely to pinpoint the time of swarming: If, for example, one has noticed colonies swarming at an apiary during the flowering of hawthorn, it is quite certain, that they will always swarm at this moment, even if the flowering itself is shifted one or two weeks earlier or later.

There have always been variations in seasonal and weather events, but they are increasing as a result of climate change. For example in 2019 the usually mild month of May was so cold in the Rhine valley that large colonies had depleted their stores and needed urgent feeding. And in January 2020, some colonies already had large surfaces of brood, the brood break had been extremely short or was absent altogether. This results in new challenges for plants, animals, and humans in their mutual interaction and especially in the natural food chains to which they are linked.

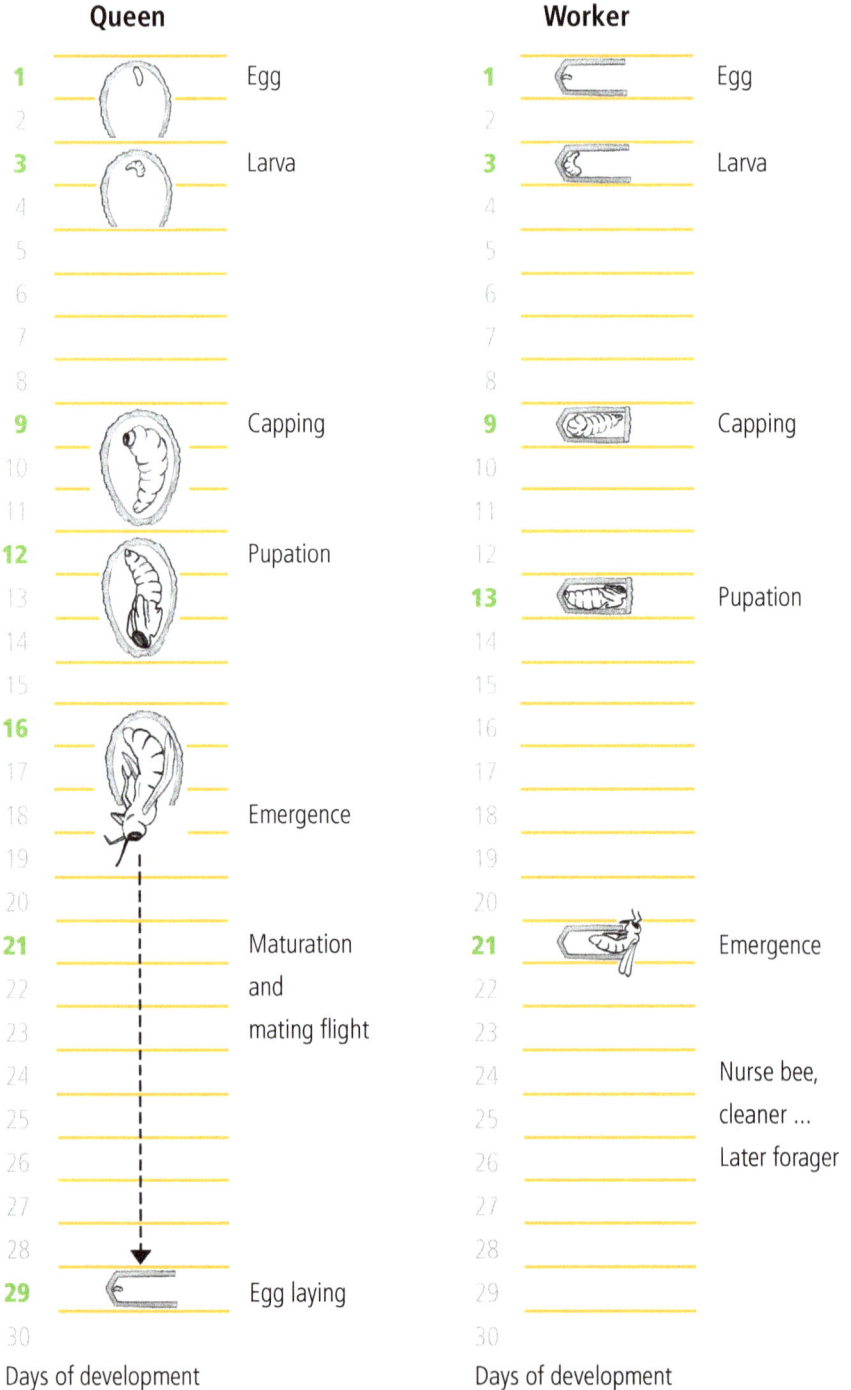

Queen

Day		Stage
1		Egg
2		
3		Larva
4		
5		
6		
7		
8		
9		Capping
10		
11		
12		Pupation
13		
14		
15		
16		
17		
18		Emergence
19		
20		
21		Maturation
22		and
23		mating flight
24		
25		
26		
27		
28		
29		Egg laying
30		

Days of development

Worker

Day		Stage
1		Egg
2		
3		Larva
4		
5		
6		
7		
8		
9		Capping
10		
11		
12		
13		Pupation
14		
15		
16		
17		
18		
19		
20		
21		Emergence
22		
23		
24		Nurse bee,
25		cleaner ...
26		Later forager
27		
28		
29		
30		

Days of development

The swarming instinct

The swarming of colonies is an expression of abundance. They are large, have generated a lot of stores and have large areas of capped brood. By contrast to the situation in early spring, when a single winter bee often has to feed a number of larvae, the situation in May is reversed: there are more young nursing bees than open brood cells. Thus it is no longer possible for all nurse bees to engage in feeding, resulting in a so called brood food congestion results. This is an important signal for the start of the swarming process. A tendency to swarm frequently occurs in large colonies with queens of two or three years of age, but it is not exceptional that even colonies with one year old queens want to swarm. Swarming is the only natural method of increasing the number of colonies, and is a climax in the life cycle of a bee colony involving the end of one colony and the beginning of a new one, i.e. dissolution and regeneration.

Three days before the primary swarm leaves the colony, major changes take place and the unity of the colony disintegrates. Foraging is reduced, or even stops altogether. The daughters change the composition of the diet for their mother, feeding less protein but more sugar. As a result, the queen loses weight, her ovaries reduce in size, and she lays fewer eggs. By contrast, the workers gain weight, because they have to ingest a lot of honey. A swarm must be able to survive for a few days without the opportunity to forage. In addition, some 80 per cent of all worker bees in a hive activate their wax glands, whereas in a colony without the urge to swarm it is only some 12 per cent. Under natural circumstances, a new home – a hole in a tree or a cavity in rocks or stone walls – needs to be quickly filled with comb.

The exodus of a swarm is a total work of art. Bees flow from the hive entrance as if a tap has been turned on. They quickly form in the air a diffuse cloud which criss-crosses the apiary in beautiful, dance like movements. The air is filled with the smell of bees, the sound is symphonic. Before its complete dispersal the cloud organises itself into visible togetherness. The first bees have already started to settle on a nearby branch, sometimes in a bush or on a fence post. As in the exodus from the Golden Hive the queen never takes the lead, she joins in the game of her daughters.

The process of swarming results in new colonies with different characteristics. The prime swarm develops quickly into a strong new colony with assured

81

offspring. Casts or secondary swarms are smaller and include an unmated queen – a major risk factor. Should the young queens fail to return to their colonies from their mating flights, the colony is doomed to die. If the weather is bad during the nuptial flight, the queen will only mate with a few drones, i.e. she will have a limited store of sperm, and her progeny will have a limited genetic diversity.

The young colony which has been left behind by the swarm has everything it needs in order to survive: combs, stores, brood – which ensures a constant supply of young bees over a longish period of time. However, it also inherits a potential load of pathogens. it will have a higher varroa burden than the newly founded colonies, and from the outset it will have to deal with the pesticides and other agrochemicals that have accumulated in its wax. It is similar to heritage among humans: it is nice to inherit the house from ones ancestors, but at times this gift includes unwelcome legacies.

Still the cloud of swarming bees is still large, soon they cluster tight together on a twig or branch and become quiet.

The different types of colonies are part of the wisdom of the Bien, they have developed and become perfected in evolutionary history.

Differences between colonies that result from swarming

Source of the bee colony	Size of the colony	Queen	Stores
Prime swarm	large	mated	none
Secondary casts	small	not mated	none
Daughter colony remaining	medium	not mated	plenty

In periods of scarce nectar in the summer months, large colonies are most at risk from starvation. Like other beekeepers, we, too have lost some of finest colonies as a result of inattention, because they finished their last drop of honey and starved. Small colonies are more able to cope with such situations, because their metabolism is much lower. The same is true for diseases. It may well be that a colony in the parent hive fails, because of the presence of spores of American or European foulbrood. And it is possible that the mite load in the prime swarm is bigger, because it develops a larger brood nest than a smaller colony originated from a secondary swarm or cast. For this reason, in autumn we sometimes do not have the heart to unite or dispose of small colonies. Time and again such 'runts' develop into strong and vibrant colonies in the following spring.

The first signs of the swarming process are so called 'play cups'. The queen lays eggs into these hemispherical cells, which are very often constructed at the lower edges of the combs on the periphery of the brood nest.

Once again it is the colony which decides whether young queens are to be raised or not. It happens frequently that the queen lays eggs, but that the bees remove them again. For reasons unknown, in such cases the colony is not in the mood for swarming after all. Beekeepers may notice this when checking for signs of swarming, when on returning to the hive after a week they still find eggs in the same play cups.

Swarm control

Whoever has the luck to place hives in her or his own garden, and is at home between 11 and 15 hours during the swarming season, can let their colonies swarm without taking any control measures. Whoever is not in this enviable situation has to check the colonies for signs of swarming every seven to nine days if she or he does not to want to miss the moment of swarming or wants to prevent uncontrolled swarming from the hive. These intervals are due to the developmental cycle of the queen, which takes nine days from an egg laid in a queen cup to the capping of the queen cell, earliest moment for a colony to issue a swarm.

Young queens are raised at lower temperatures than worker bees. This is the reason why queen cells are mostly situated at the edge of brood nest. This piece of information makes checking for signs of swarming easier, as, in order to monitor for queen cells, it is not necessary to lift all combs from the brood nest.

The queen cell has been opened by the bees on the side. An older queen, which had emerged earlier, has done away with a competitor.

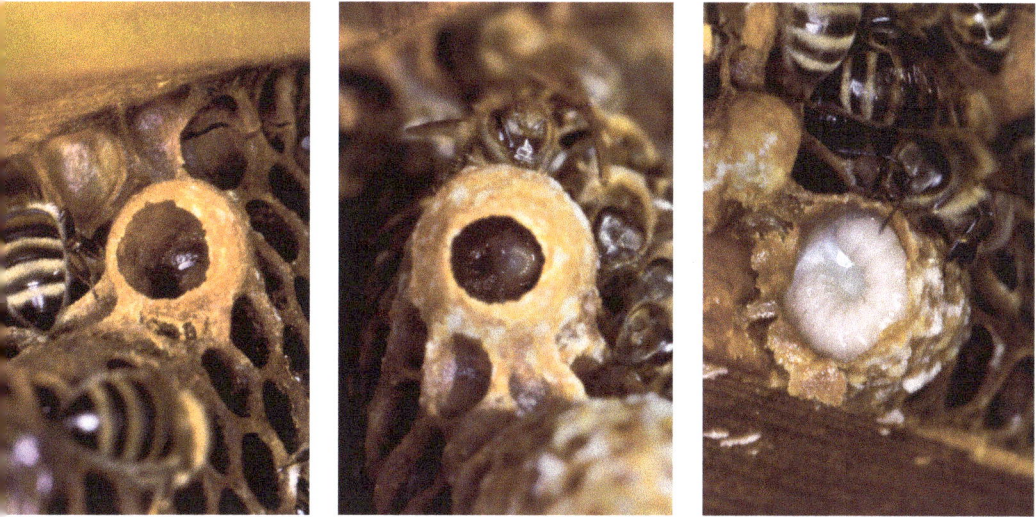

Three developmental stages of a queen: The egg has been laid (left); three days later the larva emerges (middle). The round larva (right) will soon straighten out and be capped. The queen larva is be fed with royal jelly exclusively for six days until the capping of the cell.

If the colony is ready to swarm, the eggs will hatch three days after having been laid, the larvae grow and moult for the following six days. Up to the moment of capping of the cells they will be fed exclusively and abundantly with royal jelly. In contrast to queen eggs which can be removed over and over again, the colony persists with the swarming process once the larvae have emerged. However, we have also seen bees dismantle the queen cells close to emergence, along with the larvae inside, in order to postpone swarming or, in rare instances, in order to abort the process completely. Hence the importance of close monitoring.

A colony will always raise more than one queen cell, sometimes as many as fifteen to twenty. Surplus queens will be disposed of as larvae or pupae, or they will be stung to death just prior to their emergence by older queen-sisters, which have already emerged. It is easy to distinguish cells, from which queens have emerged from those, where the queen has been killed by a sister. In the former instance a queen cell shows an opening at the bottom, in the second instance it will be opened on the side. The empty queen cells will be cleaned by the bees and after some time they will be dismantled.

Often the collection of swarms is an adventure, particularly if they are not at head height, and hanging as an easily accessible cluster on a thin branch of a tree. In order not to endanger oneself, swarms at greater height require a properly set up ladder. This will allow for a safe climbing up to the swarm and holding a swarm catching box underneath the cluster. After a knock on the twig or branch the swarm will fall heavily into it. At times swarms are located at heights that are out of reach for collection, even by means of a ladder. In such an instance a swarm catcher, a telescopic pole with a sack, helps to retrieve it. Success has also been achieved by means of a series of pieces of plastic pipe joined to each other, with a funnel at the top and with a swarm catching sack at the lower end: after a strong shake, the bees will slide into the funnel and down into the sack.

Another possibility is to place a frame with a comb and brood, or a skep, above the swarm. The bees will then congregate on the frame or follow the darkness and move inside the skep, which then allows for easy retrieval.

To collect a swarm and allow it to move in by itself

If a swarm is hanging from a branch, there is usually sufficient time to collect it into a swarm collection box, or simply into a bucket. But before doing so one should take time to observe the bees in peace and quiet. Swarms usually are very docile because they have nothing worth defending. Frequently it is possible to watch bees doing a waggle dance on the swarm cluster. These are the most experienced scout bees searching for a new home, and providing valuable information to their sisters on the cluster. Their dance contains the same information as it does for sources of nectar: direction, distance, and quality of the cavity found.

Once a swarm has been taken, the swarm collection box is placed on the ground, or the bees are poured from the bucket into the box made ready for that purpose. The box is then closed, but the entrance hole remains open. It is not possible to catch all bees. Bees on the outside of the swarm box start spreading pheromone by sticking their rear ends into the air: this smell is emitted by the Nasanov gland, named after its discoverer, and signals, together with the scent of the queen, to all the sisters still outside on the branch the direction for them to find their way into the box.

A beekeeping colleague has once told a nice story. He had been called to collect a swarm that had settled in a hedge which had grown into mesh fence.

A fair crowd of onlookers had gathered. Since it was impossible to place the swarm collection box underneath the swarm, the beekeeper very much hoped to spot the queen in order to place her in a small cage inside the swarm collection box. Fortunately shortly afterwards the queen showed up on the outside of the cluster. She must have heard his wishful thinking. As soon as the queen was in the box, all her daughters did not hesitate to join her – much to the amazement of the onlookers. The scent of the queen had attracted them like magic.

There are also situations, where the collection of a swarm is difficult or even impossible, or where a swarm absconds. It will search for a new home, and possibly move into an apiary in the vicinity. We beekeepers can remain relaxed about such an event, since we can divide the remaining colony into three or four new ones.

Only a minority of beekeepers will be able to be near their colonies during the hours around noon. And trying to find swarms in the evening after returning from work is pointless. Either the bees have already moved on, or it is almost impossible to spot them in trees, because their colour closely matches that of branches, and flight activities around the swarm will have long since ceased. In order not to lose a prime swarm, it is possible to pre-empt its issue.

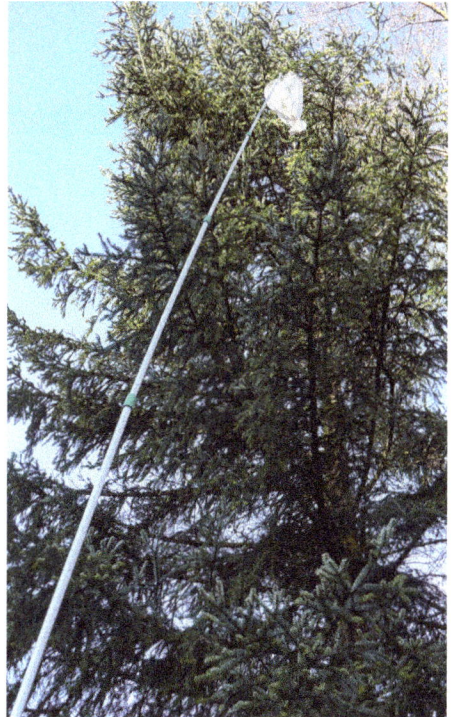

A collection sack on top of a long telescopic pole makes it possible to 'pluck' a swarm from a high tree.

87

∾ A night in the basement

A natural swarm can be hived into a Golden Hive in the same apiary on the day of its collection (see page 70); under no circumstances will the bees return to their old home. Nevertheless it is our recommendation to put the swarm in the swarm collection box for one night into a cool basement. In the darkness, the bees will form a tight cluster under the top of the box. This promotes their coming together as a colony unit. Natural swarms carry provisions for a couple of days so there is no need to feed them.

To pre-empt a swarm

If during the regular swarm monitoring there are already larvae in the queen cells (see pictures page 85), try to estimate their age and return a few days later to pre-empt the issue of the prime swarm. If you are not able to determine the developmental age of the larvae, swarm issue can be pre-empted immediately. A prime swarm will leave the colony at the earliest after the first queen cell has been capped. If queen cells are still in the egg stage, the procedure is as follows: since it cannot be known whether an egg is two, or three days old, you come back six days later, i.e. before the queen cell is capped before the dissolution of colony unity starts, as described above. Thus this method ensures that the prime swarm is taken between one and three days before capping of the queen cells. At this moment, the colony is preparing for a new beginning. Part of the bees would leave the hive together with the old queen, and those remaining will form a new colony together with a young queen.

The pre-empted prime swarm is taken to the basement; in contrast to a natural swarm it must be fed immediately, since the bees may not have had a chance to fill they stomachs with honey. Feeding is simple: either honey is smeared onto the mesh of the swarm collection box, or it is deposited on a small plate and inserted through the entrance hole of the box onto its floor. At Fischermühle we use swarm collection boxes that have mesh openings on the top through which it is possible to apply a tablespoon of honey directly into the cluster hanging there. The bees will quickly eat the honey. We recommended checking if they have finished eating it. If this is the case, a second portion should be given, and sometimes even a third portion will be required.

❧ Efficient pre-empting of prime swarm issue

If larvae have already emerged at the time of checking, try and estimate their age in order to calculate the day on which the queen cell will be capped. Depending on this, return a few days later though before the oldest queen cell has been capped and make the pre-emptive swarm. Our recommendation to beginner beekeepers who find it difficult to correctly estimate the age of a queen cell, is to do the pre-emptive swarm on the day of checking.

The frames are pulled out and checked one after the other. This requires some experience. If the queen has not been marked with a coloured dot, two tricks help: one does not 'scan' the comb, but applies a 'panorama' vision. The queen often will stand out because her hind legs are somewhat red in colour, and because her abdomen is noticeably longer than that of the worker bees. She also moves differently from her daughters. Moreover, the bees on the comb with the queen may seem agitated. After removing the dummy board, start the search with the outermost frame, and from there continue in a focused manner frame by frame. Sometimes the queen will hop from one frame to the next as if she felt chased by the beekeeper. Do not get frustrated. Even experienced beekeepers sometimes take a long time to find the queen, and may have to check all the frames two or even three times.

Once the queen has been found, she is picked up with a queen catcher purchased from a beekeeping supply company. Those with a lot of experience can catch her between thumb and index finger – she will never sting a beekeeper during this procedure. Once the queen has been caught, she is put into a cage which is then put in the warm pocket of the beekeeper's trousers.

After that, bees from five to eight combs are shaken off into a swarm collection box. It is not necessary to shake vigorously as all the bees do not have removed. The combs are returned to their original positions in the hive. If one side of a comb of the Golden Hive is covered with bees there will be some 1,300 on it. In making an artificial swarm 15,000 to 20,000 bees can easily be taken. Shake off bees only from every 2nd comb (e.g. from No. 3, 5, 7, 9, 11, 13, 15) to make sure that the pre-emptive swarm will contain bees of all different ages. At the end of the procedure free the queen and let her walk into the crowd of her daughters.

If the apiary for the new colonies is further away than the flight radius of the bees (3 km), then after one night in the basement the artificial swarm can be hived at the new location in the late afternoon. If the artificial swarm is to be returned to the same apiary in which it was prepared, it should stay in the basement for three nights. In contrast to a natural swarm, which, after issuing, would never return to the parent hive, a pre-empted swarm must 'forget' where it has been 'at home'. After three nights, bees lose the inclination to return to their old colony.

Secondary swarms (casts) and queen nucs

Under ideal weather conditions, seven days separate the primary swarm from the emergence of the first young queen. Colonies living on their own in forests rarely produce a secondary swarm as they are simply too small. But from a big colony living in a Golden Hive and ready to swarm, up to six young colonies can be made. The first one is the mother or prime swarm, the last one the colony which begins with a new queen in the old home. In between those extremes four more young colonies can be made. For this, two options exist: artificial daughter swarms and queen nucs.

However, quite frequently beekeepers with a few colonies already require neither swarms nor have they space for more hives. They can sell primary and secondary swarms through swarm exchanges, or simply sell or give them to beekeeping friends. Those without time and leisure to produce new colonies have two options to manage their colonies during the swarm season. The first is that they can produce a big primary swarm, which they then sell or give away, and then break all queen cells sparing just one or two. This will result in a colony with a new queen in the original hive. The second option is to produce an artificial swarm with the flying bees': The Golden Hive with the colony that is ready to swarm is moved a few meters away from its original site. An empty hive is put on the original site containing the old queen and some empty frames. Provided the weather is fine, all foragers will return to this new hive and join the old queen. This unit can be put in the basement where it needs to be fed, as describe above, and then given away as an artificial swarm A colony with all the house bees and many queen cells remains in the parent hive. It can be returned to its original location. In beekeeping parlance this is a 'split'. It will not swarm after this procedure, because there will not be many foragers in the colony when the young queens emerge.

❧ Pre-empted secondary swarms

Artificial secondary swarms are for experienced beekeepers. It is necessary to know the date of the emergence of the young queen. After a monitoring visit, simply write the expected date of emergence with chalk on the hive and make a note in your hive records., Then on this date check if the are close to emerging. It is quite easy to see this when looking at the lid of the cell. The bees will have substantially reduced the outside of the wax capping leaving only what looks like a thin, dark, layer similar in structure to parchment. This the queen will chew through at the time of her emergence.

It is a wonderful experience watching a queen emerging. For this the queen cells are carefully removed from the comb, carefully placed on the lid of another hive nearby, or kept in the hand. Alternatively the queens are allowed to emerge directly into a queen cage. Whichever way, the queen is temporarily housed in a queen cage, and some 10,000 bees are then shaken into the new hive, which is also stocked with five empty frames. Following this, the queen is released from the cage. into the hive This new artificial swarm will eagerly build its own combs. This procedure is obviously the closest possible one to a naturally occurring secondary swarm or afterswarm. Artificial secondary swarms must also be placed outside the flight radius of the parent colony.

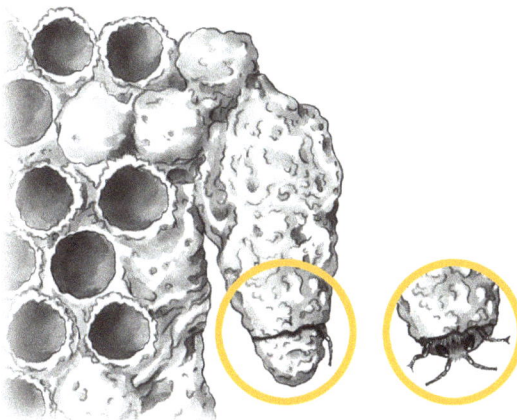

The dark line is the indicator that the queen will hatch soon.

91

Making queen nucs or splits

Queen splits with at least one queen cell can be made between the 10th and 16th day after an egg has been laid into a queen cell, in any case at the earliest after the primary swarm has left the colony. Since queen cells are sensitive to jolts in the period between the capping and the subsequent days until the hatching of the queen, they must be handled with care. It is possible to make up to four nucs from a big colony in a Golden Hive. There is no need to worry about the daughter colony remaining in the hive: there will be sufficient bees, and especially a lot of brood, usually along with plenty of honey stores.

The young queen after hatching in a nuc will need some two weeks until the beginning of egg laying. First she will fully mature, then go out on her mating flight(s), and store the sperm of many drones in her spermatheca. Sometimes the period from the egg in a queen cell to an young adult laying queen is shorter than 30 days, but it can also take longer. Even after many years of colony reproduction with this method, the period of waiting still seems to be interminably long.

By the time the young queen is ready to lay eggs, usually most of the brood will have emerged, i.e. the colony will be brood free. There are now two options for proceeding. The first is simple, but in view of the already existing combs, not quite perfect. The alternative is a bit more demanding, but perfect with respect to natural comb building.

ᐁ Making nucs or splits

A queen nuc comprises three combs. Two of these have bees and brood, which taken together will produce a colony of some 10,000 bees. There should be one or at the most two capped queen cells on them. The third comb is a storage comb. Frequently many queen cells can be found on a comb so all apart from one or two need to be removed. This is necessary in order to prevent unwanted secondary swarms. All three combs are put into an empty Golden Hive, or a comb transport box, and closed at the side with a dummy board. As there is no brood to feed, these small units will usually not forage for pollen or nectar. In addition, they are not ready to defend themselves. Thus, it is important to reduce the entrance hole to 1-2 cm2, in order to help them fend off robbers.

∾ Making a young colony from a split

The first method for making a young colony consists of placing an empty frame on each side of the split, which means that some of the combs, i.e. part of the skeleton, is taken from the mother colony. Usually, the young queens quickly start laying in these old combs. Such combs will form the centre of the brood nest, because comb that has housed brood previously, irrespective of its location in the hive, is more attractive to egg-laying queens than freshly constructed comb.

For the second method of making a young colony from a split, the procedure is as follows: as soon as the young queen starts laying eggs, she is shaken or brushed from the combs together with all the bees (her sisters!). The old combs are removed and five empty frames are placed inside the hive. The young colony will construct new combs from scratch This means that the renewal of combs will complete.

The mother colony in the old hive has all it needs: many young bees, a queen, stores, and a complete set of combs. There is no need to feed. However, if the combs are old and black, it is recommended to take away all the old combs and start the colony on fresh combs, exactly as has been described for the a queen split.

Such young colonies differ in three aspects from those of the primary swarm. Until the first new bees emerge from eggs of the new queen, the relationships between the queen and the worker bees are different. A primary swarm consists of mother and daughters In a split it is sisters and sisters. In the latter case the relational ties are not as intense as in the former.

The second aspect concerns the building of comb. In an optimal situation, a colony with a new queen in its first year will rarely, and if at all late in the season, 'think' of building drone cells. This is different from a colony with an old queen. In a new colony, drone cells, if present, will only be found at the edges of the combs. And for this reason – and this is the third aspect – colonies with young queens are fed differently from those, which result from a primary swarm.

Feeding and the expansion of the young colony

Young colonies are initially fed small portions that increase in size week by week. The reason for this is simple: if too much feed is provided initially, all cells in the newly built comb will be filled with honey. Beekeepers call this a honey bound brood nest, as a consequence the development of the colony will be slowed down. Conversely, if there is not enough food available comb building will progress only sluggishly.

As a rule of thumb, the following volumes are provided at five to seven day intervals: 0.75 litres, 1 litre, 1.5 litres, 2 litres. Feeding is then continued with 2 litre portions at the same pace, until the colony has stored the necessary winter feed. At locations with a mild climate, 15 kg of stores will suffice. In other regions 25 kg will be required. Local beekeepers may know how much feed is needed. It is always helpful to ask these friends about their experience regarding the quantities required for the winter season. Such information helps to winter colonies safely without supplementary feeding.

Colony expansion is envisaged only when the combs have been drawn out to between two thirds and three quarters of the height of the frames. Depending on the colony size, one or two empty frames are added next to the outer combs. By the end of the season all combs occupied by the bees should ideally be drawn out all the way to the bottom of the frame. This ensures that the colony has proper access to the stores during winter months.

In the cold season, the bees cluster tightly together in the space between the combs and slowly consume the honey on the margins of the winter cluster. If it is very cold, they cannot leave the cluster in which they keep each other warm. As a consequence they are prone to die because of so-called isolation starvation. This is unfortunate as other combs nearby still have plenty of stores. For this reason it is important to have an uninterrupted access to the stores.

If the comb has not been fully drawn out in the first season, the remaining space will be filled with drone brood cells rapidly in the following spring. For the hive management it is advantageous to have combs filled mostly with either worker brood cells or drone brood cells.

Large brood areas will result in a sudden growth of the colony. This dynamic of swarms should not to be underestimated and should be met with timely additions of new empty frames. If young colonies in their first year have fully drawn out more than twelve frames, the number of combs should be reduced to twelve in late summer. Frames at the margins can be removed as they are usually without brood. Empty brood cells will be filled rapidly with honey by the bees.

The combs taken out can be added back to a colony in late winter if extra stores are needed. Note that colonies have different characteristics regarding the building up of stores and their consumption.

Irrespective of the dimensions of a frame, all young colonies will initially construct a 'heart shaped' combs. These are of particular beauty in the Golden Hive. The shape is the result of the construction bees forming a 'chain' (more on this on page 15) exposed to gravity: the bees end up hanging plumb-vertical. The darker areas in this comb show that the colony already had raised brood there; the light coloured, totally white cells have not yet been touched.

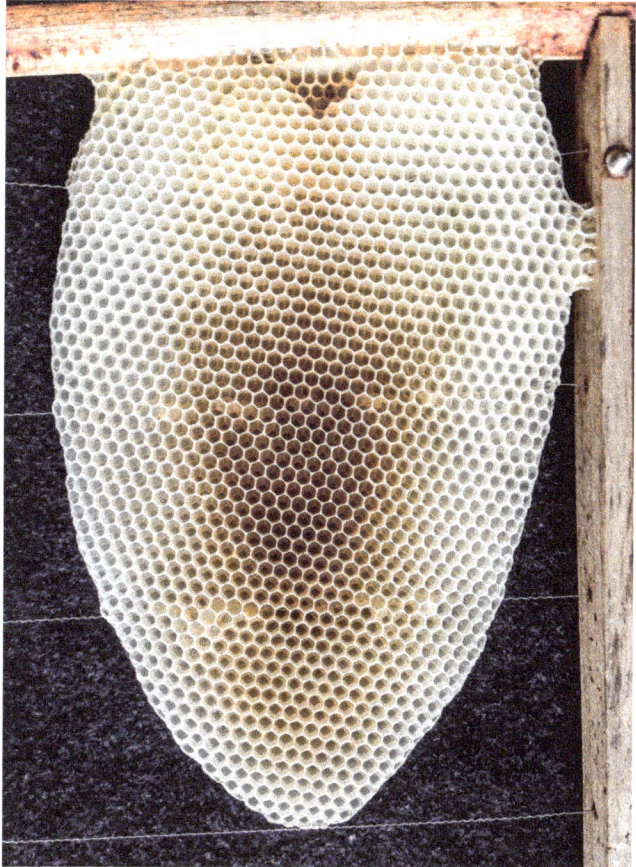

Managing comb building

In natural comb building the snow-white combs will initially have the shape of drops or of a small swarm cluster; they are called 'heart shaped' combs. At seven hundredth of a millimetre, the cell walls are incredibly thin. After a 'rough' initial construction, the bees scrape and plane them into their final shape with their mandibles.

At the latest after three to five days it is necessary to check the state of comb construction in all new colonies. Are the bees drawing the comb straight and in the middle of the frames? Any irregularities can easily be corrected with a little pressure from the hive tool, since the wax is still soft and pliable. If the bees construct a comb in a distinctly wavy shape, then it can be rotated 180°. This will cause the combs to touch each other, but the bees will quickly fix this problem by straightening out the wave and adjusting the bee space, so that they are free again to move between the combs. Checking the comb construction is important because a defect in one comb is usually continued and worsened on the following combs.

Drone cells are a welcome phenomenon in bee-appropriate beekeeping, not least because of our knowledge from wild honey bees who construct drone cells on 20 per cent of the total comb surface on average. Raising drones is a contribution to the genetic diversity of all be colonies in the region (see more on page 36).

With the construction of drone cells, colonies start to become distinctly more individualised. In natural combs, drone cells are usually located at the bottom edges or sides of the brood nest. But occasionally they can also

꩜ Regulating the construction of drone cells

Young colonies, irrespective whether they have come from a primary or a secondary swarm, should have eight to twelve fully drawn combs at the end of the summer. It is however important to begin with about five frames, and add additional ones only when two thirds of the frames have been filled on both sides of the brood nest. This makes sure that drone cells are only constructed at the bottom edges of the comb.

be found in the middle of a comb surrounded by worker cells. In such an instance, the sequence of construction will be first worker cells, then drone cells, and then worker cells again.

The construction of drone cells will vary in position and number from one colony to another. It is an expression of the unique 'character' of a colony and an expression of different sentience of each colony. Not all the drone cells will be used later on to raise drones. This is one of the reasons why natural comb building is important not only for health reasons (see more on page 33), but is also from an aesthetic and ethical perspective part of bee-friendly beekeeping.

Because a colony's comb building is highly individualised in nature and serves as its skeleton, combs from the brood nest area are never shifted around in the hive nor or even swapped into other colonies. In combs drawn on foundation, differences are less pronounced and individual characteristics reduced. The growth of naturally constructed comb will show beekeepers who have some experience certain character traits of a colony.

Drone cells are slightly bigger, and their caps are convex bulges.

Monitoring the colony population turnover

Colonies that did not swarm, will gain their maximum colony strength of up to 50,000 bees by the midsummer solstice. Underlying this impressive size is a phenomenon which beekeepers refer to as 'turnover'. The incredible egg laying capacity of a healthy queen is up to 1,500 eggs per day, more than her own body weight. Over the course of a full bee season approximately 200,000 bees are raised! This impressive vitality of the queen is made possible only because her daughters provide her with constant care and feed her with royal jelly.

The turnover, the constant renewal of the bees in a colony and the parallel growth of the overall colony strength right into summer, is part of the survival strategy which has been optimised by the honey bees over their long evolutionary history. It ensures that there are always enough young bees to do the required tasks inside the hive. At the same time it is a hygienic measure, because the high turnover of bees makes possible a quick removal of disease from the colony. As we will see, unfortunately this strategy has a counterproductive result when it comes to the varroa mite, because large brood nests favour the development of the mite population (more on this on page 111).

Large bee colonies will, depending on the region, collect nectar from cherry, dandelion, false acacia, lime and horse chestnut. In many regions this will be followed by honeydew from conifers, i.e. white fir and spruce. But in many regions, reality looks quite different. At the latest after the blossoming of the lime trees in June, the countryside within the flight radius of a bee colony may turn into a 'desert'. This is almost always the case in regions with industrial agriculture, be it arable farming or intensively managed pastures. In such a situation a beekeeper must be on high alert and monitor his colonies in order to be ready to feed if needed, otherwise it is likely that the largest and most beautiful colonies in particular would be prone to die of sudden starvation.

"In the middle of the year the colony begins to construct its winter sleigh" - this old beekeeping saying refers to the fact that bees already begin to raise the first winter bees on June 24th (St. John's day). This has to be taken into account after the honey harvest in summer.

❧ Colony management for honey harvesting

With the flowering of the goat willow the colony starts to establish a large brood nest. Thus, the number of bees will increase rapidly. If it is not already spread over 12 frames, as might be the case with a young colony from the previous year, an extra empty frame is added on both sides. It is easy to get the timing right for this. Our beekeeping friend Albert Muller has pointed out that the need for extra space is indicated by a few or even many bees sitting on the other side of the dummy board. Now is the time to enlarge the space for the colony. A third, and possibly even a fourth additional frame is best added on the side of the nest nearest hive entrance. With the first major nectar flow, frames with empty comb or foundation are added on the side away from the entrance. Similar to the expansion of space for young colonies, it is better not to expand too much in one go. It will be more economical later on to extract honey from a few completely filled honey combs rather than many half empty ones. Once two thirds of a honey comb has been capped, better even three quarters, it is ready for extraction. But no harm is done if combs are left in the hive until July, when spring and summer honeys can be extracted in one go.

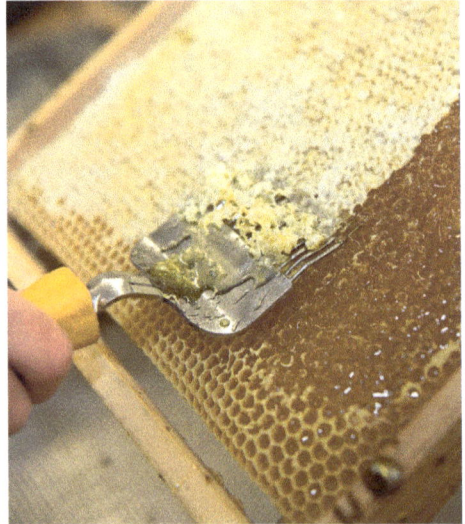

Prior to extraction the waxen lids are removed from the comb.

Centrifugal forces will cause the honey to flow out of the cells. It is necessary to adjust the rotational speed in order to avoid comb breakage.

Honey flows into a double strainer, which filters out all remnants of wax.

Keeping Varroa mites under control

Even though young colonies show low varroa mite load, because they lost many mites due to the interruption of the brood cycle, they too should be monitored for mite numbers from July onwards.

Following the honey harvest, all colonies must have their mite load checked on a monthly basis until September. It is not enough to check just one colony per apiary, because, for unknown reasons, there can be colonies with low mite counts as well as those with a very high numbers. It is also negligent to omit treatment based on the assumption that there would be only a few mites because most of them had been dealt with during winter treatment. During the warm season colonies can end up with a surprisingly high number of mites.

A key factor for mite loads is the density of colonies and the distance between the hives. In times of lack of flowers, strong colonies especially will try to rob honey from other colonies in the same or from neighbouring apiaries . Small colonies with a high mite count are not in a position to defend against such robbing. And along with the sweet honey the robbers will carry varroa mites to their homes. Some years ago the biologist Wolfgang Ritter, an expert on honey bee health, demonstrated that a mite-free colony could become infested by as many as 1,000 mites within a single week. We, too, have had similar experiences. Only three weeks after a successful treatment, which resulted in a mite drop of 1,200, a new invasion resulted in a new count of 1,000 mites in the hive. By the end of the bee season we had rid the colony of more than 5,000 mites.

❧ Danger from varroa mites

The varroa mite is a parasite which was accidentally introduced by a bee research institute into Germany at the end of the 1970s. It then spread within a few years over the world as a result of the global trade in bee colonies and queens. Australia is officially still considered to be free of mites, but even there the mite was spotted for the first time in colonies which arrived accidentally with a freighter dropping anchor in a harbour in 2019. In Asia, the mite coexists with Apis cerana, a close relative of the European honey bee, without endangering Apis cerana colonies. Because the development time of its worker bees is shorter than that of Apis mellifera workers, the mite cannot reproduce in worker brood and only with limited success in drone brood. Since drones of Apis cerana have to free themselves from their cells, they lack the strength to do so if there is a large number of varroa mites present. As a result both drone and mite die in the capped cell.

In our region the mite multiplies successfully in both worker and drone brood, and doubles in numbers every three weeks. If for example a colony overwinters with 50 mites, it will have from 3,000 of them by September.

Varroa mites depend for their reproduction on bee pupae from which they feed on their haemolymph. On adult bees, they feed on the fat body tissue, a vital organ. The weakening of the brood and the bees is made worse by the fact that the mites transmit a series of bee viruses. The Deformed Wing Virus (DWV) and the Acute Bee Paralysis Virus (ABPV) occur frequently and symptoms are easy to spot: deformed wings, and bees that can hardly move and fall off the landing board.

If a colony starts the winter period with a mite load of 3,000, it is likely that it will not survive the winter. If before the end of June there are already hundreds of mites in the colony, the disease pressure will be so high that, without treatment, the colony will die. The first step towards fighting the mite is a reliable diagnosis in order to ascertain the level of infestation (see more on page 115).

Feeding in summer and in winter

Bee colonies should be monitored for both checking the mite load and assessing the honey stores. At the beginning of August it is necessary to estimate the available food reserves. One Golden Hive frame completely capped contains at least three kilograms of honey. Depending on the region the bees should have 15 to 25 kilograms of stores by mid-September.

The fragrance of dissolved honey in sugar syrup has great attraction for all bees in the vicinity. Therefore it is important to make sure that feeding of individual colonies is done in such a way that it goes unnoticed by bees from other hives. Otherwise robbery takes place not just between two hives, but also between many or even all the hives within the apiary. Robbing usually results in the death of the colony targeted. Even strong, vital colonies can become victims. It can often be recognized right from the beginning because of intense fighting at the hive entrance. Later on the only thing noticeable is the very high number of robber bees flying in and out, far more than would be normal for ordinary foraging activities. Because they carry away large volumes of feed and honey, sticky patches can be observed around the entrance. The best way of dealing with robbing is to immediately remove the target colony to a new location outside the flight radius of three kilometres.

The best way to avoid robbing is to feed the bees in the evening, after the end of foraging activities. If more than one colony is to be fed, it is recommended to first put feeding buckets with straw or other floating aids into the hive and to cover them with a waxed cloth or a softwood particle board. Only after all colonies have been prepared in this manner is the sugar syrup quickly poured into the buckets.

If temperatures fall to less than 5 or 7°C at night, the bees will take up hardly any of the liquid winter feed. If the bees have not been fed sufficiently before winter, stress will result in the following spring when, in March and April, the bees will be short of stores. Emergency feeding is quite difficult to provide on cold April days, but can be done e.g. by providing honey or a blend of honey and fondant directly between the top bars of the frames above the area where the bees are clustered.

After the first night with below zero temperatures the colonies will start to adapt to the impending winter. As a rule of the thumb, if the weather remains

🌿 Feeding in winter

The timing for the beginning of feeding for winter depends on the availability of nectar in the respective region, and it varies significantly between various landscapes. Unlike young colonies, which are fed continuously from the start, older colonies are only given provisions for winter after the last honey harvest at the end of the foraging season late summer, often with one large portion of feed consisting of five to eight litres, with a blend of 2 parts of tea, 3 parts of sugar, and 10 per cent honey based on the amount of sugar (see page 74 for the recipe).

In a natural setting, the forager bees are busy with collecting honey stores for winter from the onset of flowering of cherry trees to the last nectar available in late summer. Thus, multiple generations of bees are involved. When we provide our bumper food supply of sugar water after the last honey harvest at the end of summer, the situation is quite different. In this instance usually only two generations of bees at the most are tasked with converting and storing the total amount of winter stores — a tremendous strain on the bees. Healthy colonies manage the task without problems. But as regards quality, the artificial feed bears no comparison with the variety of the various honeys stored over the course of a year.

Some colonies store only a small amount in their combs, even if they have emptied three buckets of feed. They have not been robbed — they simply use up the feed. The opposite can happen, too. All available cells have been filled with storage and capped. In that case it makes sense to brush the bees off the combs and to remove the combs without any bees from the hive. One then either places them into a hive that is still short of stores, or keeps them safely in a comb transport box for the feeding check towards the end of winter. A few cold days with temperatures below freezing will kill wax moths, which otherwise could cause great damage in the warmer seasons.

The amount of available stores should be noted on the hive card or in the bee diary, to make sure that hives short of supply are checked at noon late in winter (February or March) on a warm day. A notebook is a great archive to keep records with exact dates all year long, not just about the feed situation, but also for remarks on weather, vegetation, swarm related activities, varroa load and treatment, or lucky inspirations.

Three golden hives in the winter time (right), a thermal imaging
camera exhibits location and size of the winter clusters.

cool the colonies will be free of brood after 21 days, because the queen will
stop laying eggs after the first frost. From this point onwards it is possible to
reduce mites numbers with oxalic acid treatments (see more on this on page
123).

Small gaps and holes in the walls of the hives, or between the top of the
frames and the lids, will be made tight by the bees with propolis, a resin from
leaf buds of trees. In winter, the colony completely isolates itself from the
surroundings. The giant, which in summer expands on average over an area of
six kilometres in diameter, has turned into a small dwarf the size of a fist- or a
football. The bees snuggled together in the winter cluster and will only leave it
for clearing flights on the few days when temperatures are above 11°C.

Healthy colonies show their presence with a low intensity hum which can
be heard when placing one's ear against the wall of a hive. In contrast, the
sound made by small and sick colonies is louder and less uniform. If one
does not want to put one's ear against a hive, or does not own a stethoscope,
one can use a thermal camera to get an impression of the size and location of
the winter cluster at any time. If at all possible, the winter rest should not be
interrupted . We stay in touch with our colonies by remaining alert inwardly
and aware of our relationship to them.

Winter break and a new beginning

The lack of brood in a colony is a sign of the winter break, moreover its whole physiology is changed. Whereas the age reached by the sisters in summer varies between 30 and 40 days, sometimes even less, the winter bees live for six to seven months This is made possible by the tissues of the fat body which is the energy and power storage organ. In summer, it serves the nurse bees as the source of activation of the various glands producing the brood food, but in winter it remains mostly untouched. A hormonal change occurs in parallel with the lack of use of the fat body tissue, both within individual bees and in the whole colony. The concentration of juvenile hormone, which in summer triggers the nursing activities, drops to 20 per cent. Not unexpectedly the change also triggers a reaction by the varroa mites. We do not know whether it is the lack of brood on which the mite depends for its reproduction to change, or whether the composition of fat body tissue causes a transformation into a winter mite. In any case the parasites fix themselves tightly to their hosts' bodies. The phenomenon is noticeable in the varroa drop count: In winter, one mite per day on the monitoring board is equivalent to approximately 400 mites in the colony.

From the middle of January onwards the colony will have already begun, unnoticed in the same manner in which it begins in July to raise winter bees, to care for a few initial small patches of brood. These will be expanded only at the onset of flowering of the goat willow. The following weeks represent a tremendous strain for the colony. Because, with the expansion of the brood nest the winter bees, now decreasing in number, have to feed, warm, and nurse a growing number of larvae, the ratio of larvae to nurse bees frequently reaches 3:1. In only a few weeks this ratio is begins to reverse, initially 1:1, then 1:2, and later still there are so many young bees that not all of them can be involved in brood care. This results in brood food congestion in their glands, which is one of the triggers for swarming. The colony exhibits strength and vitality. It is the beginning of the new bee year!

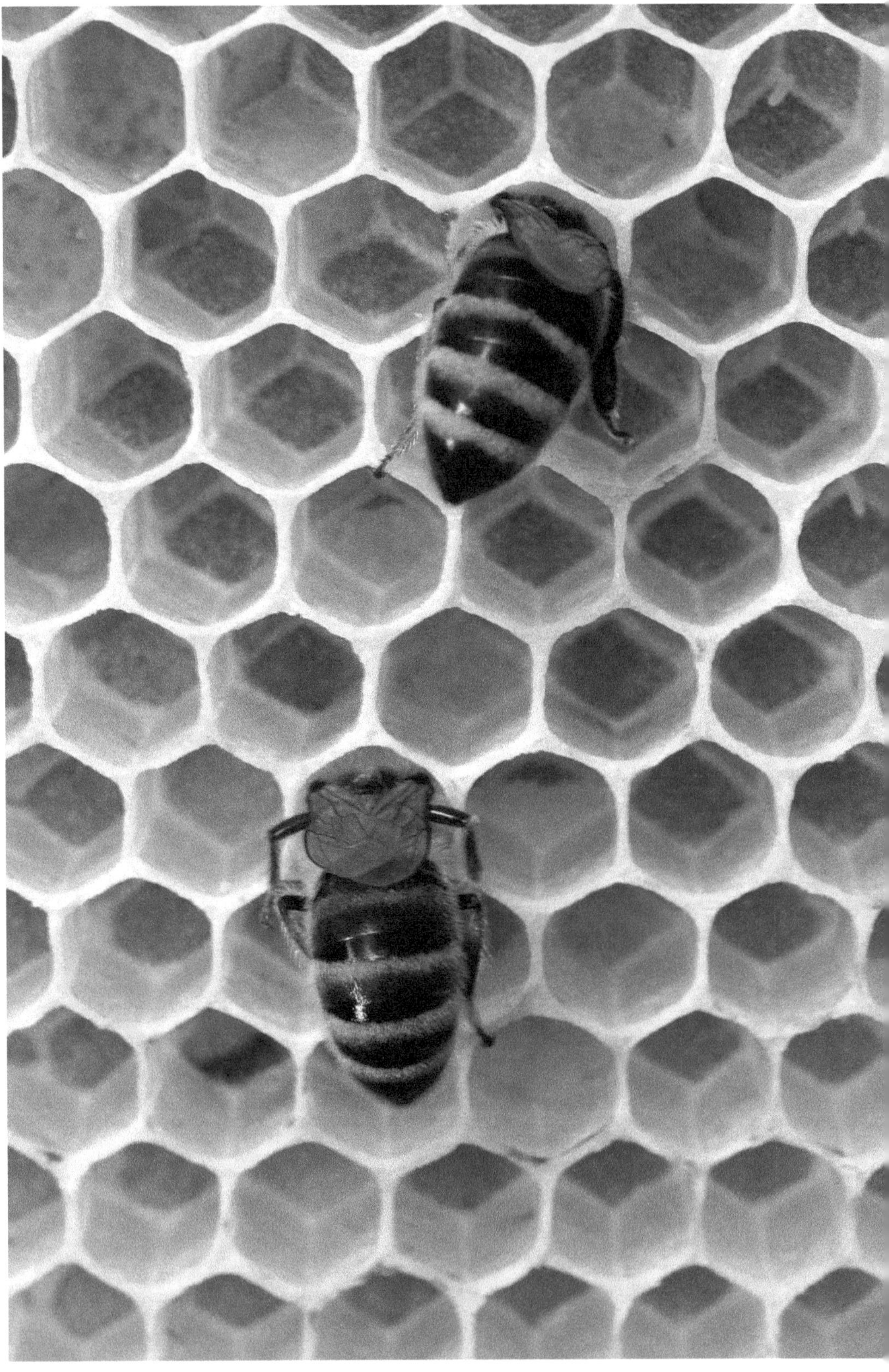

Bee health – bee diseases

Anyone who has shared the lives of relatives or close friends in situations of serious illness, has possibly also experienced how a life may have changed into some other fulfilling existence, despite the disease. It may seem in such instances that an illness represents not just a threat, which has to be dealt with as quickly and as efficiently as possible, but that it can lead to a new life plan that may somehow be meaningful in its own way.

When it comes to bees, it would be wrong to search for the meaning of disease on any psychological or mental level, but disease also has meaning in the animal world. It plays the important role of providing the continuing opportunity for the further development and adaptation of an organism's defences, e.g. its immune system and specific behaviours. On the one hand, parasites, pathogenic bacteria and viruses cause a weakening, but on the other hand they can also be viewed as 'engines' for development and evolution. Seen from this perspective, it comes as no surprise that all bee colonies always carry within them a large number of pathogens – bacteria, single-cell organisms, fungi and viruses, which in most instances do not cause the majority of colonies to fall ill.

This holds true for the dreaded and notifiable American foulbrood, European foulbrood, which is common in Switzerland and which is also known as 'sourbrood', Nosema (spring dysentery), and a multiplicity of bee viruses such as acute bee paralysis virus, the sacbrood virus, and, last but not least, the parasitic varroa mite, which on the one hand harms the bees directly, and on the other acts as a vector for the deformed wing virus and a series of other viruses (for more on this see page 103).

A disease will only erupt when an illness causing agent is present and if the animal, in this instance the bee, has a disposition for it or suffers from impaired health. From this perspective, i.e. considered biologically, artificially fighting disease is problematic, because the affected organisms do not thereby 'learn' to defend themselves. Furthermore, parasites and pathogens are surprisingly quick in their resistance development, as demonstrated by the varroa mite's resistance against many chemical and synthetic treatments, which were used in the early days of the 'varroa era'.

Wolfgang Ritter, the former director of the department for bee research at Freiburg Institute for Animal Hygiene, Germany, admits in a self-critical assessment that it was a mistake in the late 1970s to fight the varroa mite with chemical compounds. He is convinced that without that intervention bee colonies today would have long ago developed mechanisms of tolerance and resistance.

And events have proven him correct. Beekeepers in many regions of Europe have stopped treating their colonies against the varroa mite and in do not suffer substantial colony losses in the long term.

A few things are worth mentioning in this connection. Tolerance and resistance traits seem to develop 'by themselves' after approximately four to six years, provided that colonies are left untreated in order to let them develop these traits. This is the case with the scientifically well documented colonies on Gotland and in Avignon, and is confirmed by beekeepers in Norway, Switzerland, Wales, and the Netherlands, all of whom are no longer treating against varroa. This is in contrast to the more than 20 years of breeding efforts by private and public research institutions, which until now have achieved little if any progress in their efforts to breed for varroa resistance. Apparently, characteristics of resistance and tolerance elude traditional breeding strategies of simple inheritable traits based a single or just a few hereditary factors.

On the other hand – and this is the confirmation for the assumption outlined above – there are a number of known factors which favour tolerance and resistance. The first is a direct link to the beekeeping method. Secondly, a low density of colonies reduces the transmission of mites from one colony to another. This so-called horizontal transmission entails two problems. The first is that colonies which have already developed a level of tolerance and resistance, e.g. by removing mites, are threatened again and again by newly introduced mites. The second has to do with the horizontal transmission. In this the virulence of a parasite increases, i.e. the mites become more harmful. Perhaps this is the reason why in the 1980s colonies with 10,000 mites survived the winter, whereas toady they are already dead by December with only 3,000 parasites.

Another characteristic of tolerant and resistant colonies is their size. They are usually smaller than the standard colonies kept by beekeepers for honey production, and they rear less brood as well as swarm more frequently. Unfor-

All over the globe colonies survive in coexistence with the varroa mite (marked with blue dots). They are found on all continents and their respective climates, belong to different sub-species, live under highly divergent management practices and colony densities. Obviously the capacity to adapt, which enabled the honey bee to survive for 50 million years, remains intact even today. Up to now, there is not a single instance in which the resistance or tolerance to varroa is the result of targeted, artificial breeding. (Illustration based on Barbara Locke, 2015; expanded).

tunately, science seems to show little interest in such colonies, even though they, too, can in favourable forage conditions produce enough honey to make a harvest possible, albeit modest.

In a natural setting, colonies which in terms of size and honey production do not achieve maximum results, are still perfectly capable of fulfilling their pollination tasks. For a blossom it does not matter if the bee lives in a large or a small colony.

We should not forget that the so called ecosystem service of honeybees, i.e. pollination, outstrips the economic value of honey by almost a factor of twenty!

Effects on varroa infestation

A look at the reproductive cycle of varroa reveals that the mite is actually not at all well adapted to the honey bee – despite its devastating impact. Just prior to capping of a brood cell on day 8 or 9, a fertilized female mite crawls in to the cell. She does not lay any eggs in the first 70 hours, as her gametes – sperm and eggs – have not yet fully matured. Afterwards a fertilized egg is laid every 30 hours. The first egg will always become a male, all others develop into females. The female mites will be inseminated by their brother inside the brood cell. The breeding success rate seems to be modest. On average 1.3 fertilized mites will emerge from one worker bee cell. This means that on average only for one of three "mother-mites" a fertile daughter hatches. This modest success rate, however, is sufficient to lead to a doubling of the number of mites every three to four weeks, or an increase from 50 mites at the beginning of the bee season to 3,000 in late summer.

Female mites prefer drone brood over worker brood by a factor of eight, provided it is available. Because drone brood remains capped for a longer period, the reproduction success rate, with 1.8 or 2 mites per reproduction cycle, is markedly higher than in worker brood.

One can speculate about character and behavioural traits which may benefit resistance or tolerance of bee colonies: for example the number of offspring would be reduced if the egg laying of the female mites is delayed. This phenomenon is called suppression of mite reproduction (SMR). It is also possible to imagine that the bees clear infected brood cells and groom each other in order to get the mites off their bodies. Such behaviour has actually been observed. In the case of the varroa resistant colonies in Gotland, the larvae produce a substance which delays the oviposition of the mite. Colonies in Avignon clear out infected bee brood. This behaviour is known as varroa sensitive hygiene (VSH). And beekeepers examining through a magnifying glass dead mites from the varroa board have frequently noted bite marks: antennae or legs are missing, the carapaces show signs of chewing. It is assumed that such damage is inflicted when bees try to get rid of mites from each other, a behaviour known as grooming.

Finally video recordings of the brood nest have shown that bees open and re-close capped cells often up to twelve times – known as 'uncapping-recap-

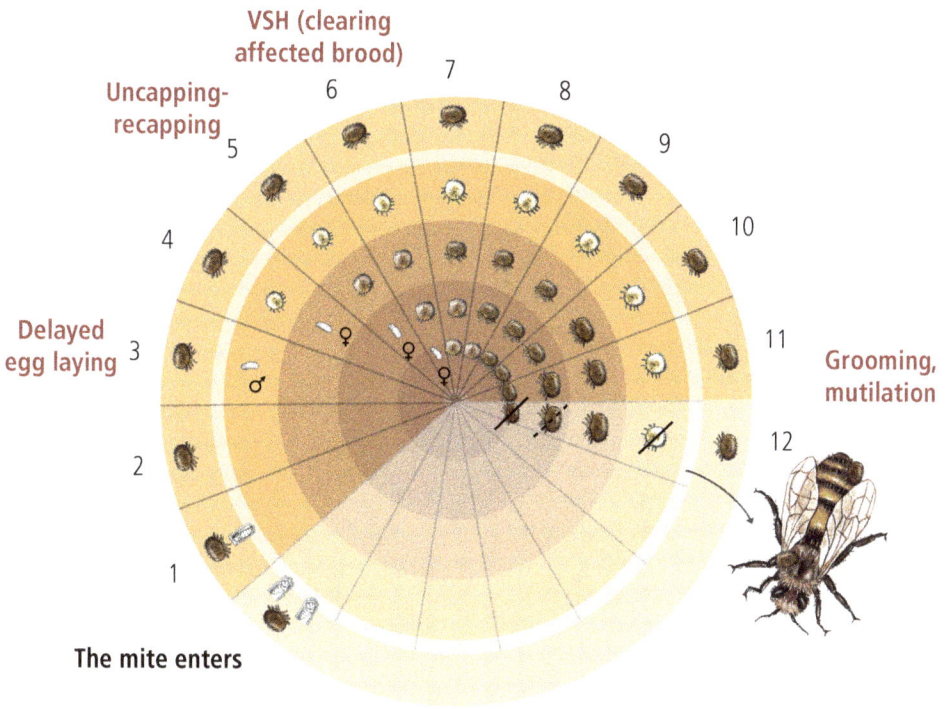

The reproductive cycle of the mite is shown in a schematic manner. The female Varroa mite enters a brood cell shortly before it is capped and emerges at the same time as the worker bee with one, rarely two fully matured daughters. The bees fight off the intruders with four behavioural traits shown by varroa resistant colonies (from a delayed laying of eggs to grooming). It is known that all colonies exert these characteristics in varying intensities.

ping'. It has been assumed that this behaviour disrupts the development of the mite. It may be compared to the situation of a couple sleeping in one bed, where one partner keeps pulling the blanket on his/her side. One partner wakes up, pulls the blanket back to his/her side, but now the other partner wakes up and the cycle is repeated … restful sleep becomes impossible.

Experts estimate that a reduction of the reproduction rate by a mere 30 per cent would suffice to ensure the survival of the colonies, and thus coexistence with the mite.

While we applaud the efforts of our colleagues, and are involved in our own research projects into issues of tolerance, we strongly advocate treating the colonies against varroa mites. Otherwise it can easily happen that small-scale beekeepers with three or four colonies may end up with a total loss as soon as after their first winter, and almost certainly after the second winter. Beekeeping beginners in particular should be enabled to enjoy the successful wintering of their colonies. On top of that, all beekeepers share responsibilities towards their own bees and colonies in the neighbourhood. And finally we also have to point out that, at least in Germany, treatment against varroa mites is prescribed by law.

At the same time we would like to encourage experienced beekeepers to consult with bee research institutes and take steps towards developing tolerant and resistant colonies. Colonies that look promising can, after careful monitoring, be selected and multiplied on a local level. In our opinion the only way to achieve genuine success is a joint effort involving a regionally co-ordinated avoidance of varroa treatment, which can lead to a treatment-free, locally adapted landrace.

The field trial on the Swedish island Gotland in the Baltic Sea, mentioned previously, provides an illustration of what it is about. This international research project involved the Swedish bee researcher Ingmar Fries, the biologist Peter Rosenkranz of the Institute for Bee Research in the German Federal State of Baden-Württemberg and researchers from the Centre for Bee Research in Switzerland. 150 colonies were placed in several apiaries in a restricted region of the island. For a period of seven years, the researchers examined the impact of the varroa infestation on the bee population. Out of the 150 colonies from the onset of the research project, only six, i.e. less than five per cent, were still alive after three years without treatment. However, the winter losses became fewer over the following years, and the bees better managed to coexist with the varroa infestation. Nevertheless, according to an article in the Swiss Beekeeping Journal the remaining colonies in the project had to be treated with oxalic acid in 2019, as otherwise their survival over the winter season would have been in serious doubt.

Varroa diagnosis

All beekeepers should make it their duty to count the mite burdens of their colonies monthly from July to the end of September. We have a rather low opinion of applying treatments in small apiaries without a preceding diagnosis. The situation is different for professional beekeepers who work with hundreds or thousands of colonies. They do not have the time and leisure to differentiate between varying varroa burdens in individual colonies and hence follow a prescriptive routine: in most instances they use formic acid and mostly oxalic acid in every colony.

Two methods have proved reliable for to checking the mite count in a colony. The floor board shows the natural mite drop, the icing sugar method actively removes mites from bees so that they can be counted.

For both methods the rule stands: 5 mites per day in July are unproblematic. The corresponding number for August is 10, and the same holds true for September. There are experts who claim that a count of 15 mites is too high in this month. If this threshold is exceeded, a treatment in the next two to three weeks is necessary. Irrespective of the month, if the mite counts are 25 or more, immediate treatment is recommended (see for this page 119).

Mite diagnosis and need for action

Method of diagnosis and need for action	July	August	September
Mite count on the floor Board, treat within three weeks	5–25 mites	10–25 mites	15–25 mites
Icing sugar method, treat within three weeks	5–25 mites	10–25 mites	10–25 mites
Damage threshold, treat immediately	25 mites	25 mites	25 mites

Note: During winter the mite drop should not exceed 0.5 to 1.0 mites per day, otherwise the colonies will encounter problems at the end of the cold season.

❧ Analysing the bottom board

This method, where a clean bottom board, or, in the latest models of the Golden Hive, two shallow trays for the diagnosis of varroa, is inserted via the monitoring hatch allows counting the natural mite drop without having to open to hive. After two to three days, all dead mites that have fallen off naturally are counted, and the number is divided by the number of days since the insertion. Anyone with spare time might want to check with a lens whether some mites have their legs missing/bitten off, or show other damage. If so, this may point to grooming (see page 112).

It is important, that neither bees nor ants access the mites, because both would remove them from the counting surface. The ants remove them, because they are a welcome source of food, the bees do so in their attempt to keep the hive clean and tidy. Removal by ants can be hindered by putting sheets of kitchen paper towel soaked with an edible oil onto the counting surface. The removal of mites by the bees is suppressed by metal meshes or grids placed on top of the counting surface.

∾ The icing sugar method

Icing sugar is the second method to reliably count the mite infestation. It has been developed by the bee research institute in Kirchhain (Germany). The first frame without brood is slowly pulled from the hive, and the bees are shaken onto a sheet or a piece of thick paper (size A3). A pot of the type used to collect urine samples (100 ml) is filled with bees, and they are quickly transferred into a one litre container. It is then closed with a mesh-lid, which prevents the bees from getting out. This cup can either be purchased, or made oneself.

Three tablespoons of fine, dry icing sugar are added to the bees in the container through the mesh, and the bee-sugar mixture is then gently shaken for some three minutes, by which time all the bees should be nicely covered in sugar. Next the container is turned onto its head, and the icing sugar is transferred by means of strong knocks into the fine sieve (e.g. from the double set used for straining honey). The bees can now be returned to the colony, where they will be quickly cleaned by their sisters. The sugar falls through the mesh of the sieve but the mites remain in it and can be counted. Counting is made even easier when pouring the mites onto a white surface where the remaining contents of the sieve can be spread out. This treatment may look rough, but studies have shown that the bees suffer no damage.

Why do we use the first frame without brood? The mites prefer to sit on young nurse bees, who are usually on cells with open brood, and from them they jump into the brood cells. Because of that, the number of mites on these bees is higher than on those on a comb without brood.

Research undertaken in Switzerland has demonstrated that the average of two counts per colony results in reliable results. Because we are doing monthly counts from July to September, we only need to repeat a count if in doubt or in instances of unrealistically high mite counts.

Ralph Büchler of the Kirchhain Bee Research Institute in Germany has estimated that a summer count of one mite/day either on the bottom floor or found with the icing sugar method represents approximately 150-250 mites in the whole colony. In experiments in or apiary we have found a lower number: only some 70-80 mites. In winter one mite/day represents about 400 mites in the colony (for more on this see page 107).

It is our opinion that these numbers should be double checked with more data. We are happy to receive numbers from natural mite drops or from icing sugar checks (see address on page 168). Since every therapy relies on a proper diagnosis, it would be useful to know more about the correlation between the natural drop of dead mites and the number of mites in a colony.

Treatments should not be applied without a careful mite count, similarly to not prescribing antibiotics for to humans without a prior medical examination. Some regional beekeeping associations nowadays recommend simultaneous treatments in their geographical area, which to us seems to be the equivalent of giving prophylactic chemotherapy to all people in an area without having made a prior disease assessment. This drastic image was not suggested by 'treatment refusers', but by Richard Wyss, the former president of the Swiss Beekeepers Association, BienenSchweiz.

We share Ralph Büchler's opinion, that only the permanent presence of mites – in particular during the winter months – provides the stimulus for colonies to constantly show hygienic behaviour – a precondition for tolerance and resistance. We must learn to treat neither too soon nor too late.

But: "Be prepared for tears", is a comment from our friend, David Heaf, an experienced beekeeper and author of several books on the Warré Hive and treatment free beekeeping in Wales, who is one of the first there to have stopped treatment, and who is part of a treatment-free trend in his home region. Indeed, in one early year of treatment-free only four of his 12 colonies survived winter. But from 2011 to 2020 his average winter losses are 8.7%. As he allows swarming and also collects swarms in bait hives, he is always able to replace his winter losses.

Varroa treatment

Irrespective of the type of hive in use there are nowadays three different methods for reducing the number of mites in colonies managed in a bee-appropriate manner: physical, chemical and biological.

Physical method

Hyperthermia is a physical method which was tested some years ago at Hohenheim University in Germany, not long after the varroa mite had been inadvertently introduced into this country. Bees but not the varroa mite survive well at a temperature of 42°C and optimal humidity. In practice it soon became apparent that the bees ventilate the heat out of the hive before it reaches the brood and the mites. Only after heat was applied directly to brood combs free of bees was the heat treatment effective. However, at the time of writing there is no satisfactory equipment available for heat-based treatment of the Golden Hive.

Chemical method

The most common treatments nowadays are chemical. Medicines for treating bees have to be approved for use by the government authorities. Because the varroa mites have developed resistance to many synthetic compounds, formic and oxalic acids are the substances of choice. Under a bee-appropriate management system, formic acid is used for summer treatments, because it kills mites both on the bees and in the brood. On the other hand, oxalic acid reaches only the mites on bees, which is why it is used to deal with remaining mites in the brood-free season.

The rate of evaporation of formic acid depends on the temperature and moisture content of the air. A 95 per cent efficiency is achieved between 18°and 30°C at low levels of humidity.

The treatment causes significant damage to uncapped brood. There are two methods of application: shock and long term treatment. We prefer the long term treatment, because it results in less collateral damage for the bees and the queen compared to the shock method. Formic acid is legally approved as a 60 per cent solution.

Die Stoßbehandlung wenden wir nur in zwei Fällen an: Erstens, wenn die

❧ Long term treatment with formic acid

When applying the long term treatment in the Golden Hive, we use the 'Nassenheider Evaporator Universal H', which is fixed into an empty frame. For this purpose we have added an additional horizontal bar approximately 15 cm above the bottom bar of a frame, from which a fleece is stretched and fixed to the top bar. Onto this is fixed the Nassenheider Evaporator. During the treatment, the ventilation openings in the floor as well as the opening behind the dummy board on top should be kept closed.

The Nassenheider Evaporator is assembled according to the instructions that come with it. The tank is filled with 180 ml of a 60 per cent solution of formic acid.

The Nassenheider Diffuser is mounted into a frame and placed between the brood nest (frames on the right side) and the food area (frames on the left side). The formic acid blends into the air inside the colony via the fleece fixed to the top of the frame. Feeding is being continued after the treatment via the feed bucket.

When doing this, protective gloves and safety goggles should be worn because formic acid is strongly corrosive. The acid drips onto the inserted fleece via a wick. A plastic sheet below the fleece prevents the acid from soaking into the wood. The acid rises in the fleece and in doing so evaporates into the air inside the hive.

In a Golden Hive the evaporator is placed between the first

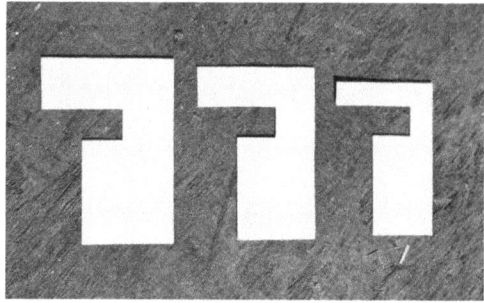

The wick for the Nassenheider Diffuser is available in three sizes. The larger it is the more formic acid will drip onto the fleece. From there the acid is rising and diffusing into the air inside the hive.

two brood-free frames, on the side of the nest that is furthest away from the entrance. The rate of evaporation should be 12 to 15 ml/day. Proper regulation of evaporation is achieved by choosing the correct wick. Treatment should last at least 12 days. It is necessary to check the rate of evaporation after three days. The evaporator tank is marked with a scale that makes it easy to check volume evaporated per day. If the quantity is less than 12 to15 ml/day, then a larger wick needs to be inserted. If it is more, it is necessary to top up the formic acid and to use a smaller wick. The differently sized wicks are part of the evaporator package, and they can also easily be purchased as spares.

Some colleagues with Golden Hives operate the 'Nassenheider Evaporator Professional' horizontally by putting it on a fleece which has been spread in a tray. But in order to do it this way, an eke of 10 cm height is placed on top of the hive. The formic acid drops onto the fleece spread out on the tray. Fill levels and duration of diffusion are the same as for the vertical method of application..

We apply the shock treatment only in two instances: the first is when temperatures during the day are above 30°C, but fall below 25°C during the night. The second is when colonies have constructed a fixed comb system, where the bees easily block the gaps between the combs against the acidified air, thus preventing a long term treatment from being effective.

For a shock treatment, 30 ml of a 60 per cent solution of formic acid are poured onto an absorbent cloth, which is placed onto a plastic film and then pushed underneath the brood nest. This treatment is repeated three times at intervals of two to four days. This time-consuming method can also rid colonies of the mites.

Sometimes the queen stops laying eggs after a treatment with formic acid in which case the affected colony needs checking at shorter intervals. In most instances the queen will resume egg laying, but if she does not do so by the time two weeks have passed since the treatment, the queen needs to be replaced or the colony needs to be terminated.

Some beekeepers have stopped formic acid treatments altogether because of the serious side effects. Instead they apply oxalic acid in the summer. A 3.5 per cent solution for spraying (trade name Oxuvar® 5.7%) is applied three to four times at four day intervals onto the bees on the combs (for details of this application see page 124).

This application also requires a lot of time. Because each female mite does not immediately enter a new brood cell after its birth, but instead stays on a bee for at least four days through what is called the phoretic phase, it can be sprayed with oxalic acid during this period. Since worker brood is capped for twelve days, it is necessary to treat at least three times. If capped drone brood remains, a fourth treatment is recommended.

An alternative is the sublimation of oxalic acid. Special tools for this are available, e.g. Varrox or Oxomat (for this see page 124).

Winter treatment – removing residual mites from the colonies

Following the first night with temperatures staying below 0°C, the queen stops laying eggs. Unless warmer winter days followed the frost, 21 days later the colony will be brood free. From this point onward it is possible to remove the residual mite population with oxalic acid, which is available ready made for veterinary purposes (on prescription) in pharmacies or from beekeeping supply companies. In Germany two methods of application are licensed: trickling or spraying. In Austria and in Switzerland it is also legal to apply oxalic acid by sublimation. If applied correctly, the effectiveness of all three methods is at least 95 per cent. We recommend treatment if in a brood-free colony from the end of October onwards more than one mite dropped per day is recorded. Ideally, a colony should have no more than 50 mites at the beginning of the winter season.

We repeat our mantra: Count the mites! If following a treatment with oxalic acid 1,000 or more mites drop, a second treatment is necessary, because there will still be more than 50 mites in the colony. This would mean that in the new year, prior to the honey harvest in July, more than 3,000 mites will be in the hive thereby threatening the survival of the colony.

Acid treatments are permitted only after the honey harvest. In warm winters, when the bees sometimes continue raising brood without a break, prior to the treatment we sometimes cut out any brood patches present with a knife.

When applying oxalic acid by dripping, no second treatment of the winter bees must take place, as the sugary solution is ingested by the bees and results in serious damage to the digestive tracts. The risk is high that a second treatment will results in the colony death. When applying oxalic acid by spraying or sublimation, more than one treatment is possible.

Treatment by spraying should only be done if the temperatures at noon reach at least 10°C. Winters in recent years have always had days with lunchtime temperatures that allowed treatments.

🐝 Removing residual mites with oxalic acid

All three methods of application, trickling, spraying or sublimation, have the same degree or effectiveness of at least 95 per cent if applied correctly.

If the oxalic acid is applied by **trickling**, then a 3.5 per cent sugary solution is applied. The licensed product consists of two components, oxalic acid dihydrate solution and sugar (sucrose), which are mixed just prior to the application. Using a syringe, approximately 5 ml per gap between two frames are dripped onto the bees. A torch/flashlight will help to see where exactly the bees are grouped. The treatment has its best efficacy when the bees are together in a cluster.

The spraying method is done with a 3.5 per cent oxalic acid solution (without sugar). Just prior to application the oxalic acid dihydrate solution is diluted with water according to the instructions of the manufacturer. The solution can be applied from the middle to the end of December in temperatures from 6-11°C. If there are many bees flying, efficacy will be reduced. The frames are lifted out one by one, and sprayed with a spray can or bottle on both sides with at most 5 ml, i.e. five squirts, of a fine mist. Those frames which have been treated can be covered with the wax cloth. To finish, the dummy board is pushed firmly back into its original location. When working with Golden Hives, an elegant solution is to link the spray can or bottle via a longer silicone tube to a bigger container. At the beginning of the treatment, all frames are pushed to the opposite side of the hive, and after spraying individually on both sides returned to their original locations. Protective measures must not be neglected (see opposite page).

When using the sublimation method, depending on the size of the colony 1-2 g of oxalic acid (tablets or powder, which can be measured exactly with a measuring spoon) are put onto the Varrox-evaporator that is then pushed through the entrance underneath the winter cluster and the remaining gaps in the entrance are is blocked with foam plastic. After a short while acidic vapour will escape through cracks in the hive. This will cease after approximately three minutes when sublimation has come to an end. The apparatus is withdrawn and the entrance hole closed completely for another ten minutes.

All three methods result in mites dropping over a period of two weeks after treatment. Only at the end of this period is a reliable mite count possible. Take the time for that as it helps the decision whether or not to give a second treatment. It helps to count mites every three to five days, noting the results, as otherwise the presence of other hive debris will make counting more difficult.

Please be aware in all three methods: formic and oxalic acid are organic acids, which were originally discovered in ants and plants. They are chemically produced on an industrial scale as treatments for bees. They are very corrosive and therefore have to be stored safely. When applying them, the manufacturer's guidelines should be followed. Use rubber gloves, protective goggles, and a protective mask for mouth and nose. When dealing with these acids a container of tap water should always be kept handy at the apiary, for rinsing or dilution in case of an accident, or simply for when something needs to be cleaned up, e.g. in case spillage when filling a container.

The sublimation of oxalic acid has been legalised in Austria and Switzerland for many years. For this method 2 g of oxalic acid dihydrate powder or tablets are inserted into the Golden Hive on the spoon of a sublimator, with the entrance hole blocked. By means of an electric or a gas heater the oxalic acid crystals are heated to the point that they sublime. The resulting vapour penetrates all gaps between all the combs and gets into all corners of the hive. If possible the heating device is placed underneath the combs where the bees are clustered. The Varrox sublimation tool was developed specifically for use in our own apiary Fischermühle. There are other tools available but their effectiveness may be different.

Frequently health warnings are issued in particular with respect to treatment with the oxalic acid sublimation method. No concerns have been raised against this method of treatment in countries which allow sublimation. This is of course conditional on following proper application procedures as described in the instructions that are supplied with the Varrox Vaporiser. Research by Tübingen University Institute for Occupational and Social Medicine in Germany has not found increased levels of hazard when applying oxalic acid by any of the three methods described here.

Biological methods of treatment

In what follows we would like to describe three biological methods developed and reported in detail by Ralph Büchler and colleagues in the Bee Research Institute in Kirchhain. Information leaflets are free to download via the internet (see address on page 171).

By contrast to the formic acid method, which is applied in July or August, or the oxalic acid method, which is only applied when there is no brood, usually at the end of November or early in December, the three biological methods can be implemented from May until early in August. For animal-ethical reasons, our beekeeper colleague Michel Colette limits application of the methods to the end of June when the swarming period has definitely ended. All methods share a common characteristic: they imitate something which happens in every instance of swarming, namely that colonies pass through a period free of brood.

Total removal of brood

This method is the easiest to apply of all three, and also suitable for beginners. All the bees and the queen are shaken or brushed from the frames into an empty hive, which is then put at the same place as the parent colony. Provided it is well fed, this 'naked' swarm will quickly construct new combs. Cautious beekeepers treat such a colony after some six days, i.e. before the first brood cells are capped, by spraying oxalic acid (see page 124). By this method most of the mites, which may have come into the artificial swarm on the bees, are removed. A total removal of brood should not be done after the middle of July, because it will then be too late for the bees to construct sufficient natural comb. and therefore not enough winter bees can be raised, and there will be too little space.

There are two methods of dealing with the combs which contain not only mites but also brood and stores. Either they removed and rendered, or used to make a split. In the latter instance, one comb covered with bees is left in the parent hive. This number is sufficient for looking after the brood, particularly since in the subsequent days almost 1,000 additional bees will hatch daily. As most of these have never yet left their hive, they will not return to their old queen. In these queen-free colonies worker larvae will quickly be transformed into emergency queens, which, after their mating flights, will quickly start laying eggs. A period free of brood thus is ensured here, too, and the colony with its young queen has a reduced load of mites. Before capping of the first brood, it is possible to apply oxalic acid to remove remaining mites.

Both colony types can often manage the winter months without further treatments, provided there are sufficient stores. However, it is our recommendation to start assessing the mite load again at the latest in September. R. Büchler and his colleagues have found that colonies resulting from such drastic interventions are nevertheless comparable in size those treated with formic acid in summer.

For a total brood removal the bees are brushed from the combs. The brood combs are removed and empty frames are substituted in their place. The bees will construct fresh combs immediately. Since the colony is free from brood varroa mites on the bees can be removed by oxalic acid.

The comb trapping method

This method, which is rather time consuming and complex, is applied from May to the end of July. It requires a comb-cage, which is constructed with walls made from queen excluders, and an empty comb for the trap, i.e. an old brood-free comb.

The queen is transferred into the comb-cage which is then returned to the colony. Her daughters will quickly start feeding and grooming her. Should the comb be completely drawn to the frame walls, a hole needs to be cut into it, such that queen and bees can easily access to both sides. At a rate of 1,500 eggs per day, all the cells in the comb will contain eggs in six days. At a rate of 1,000 eggs per day this will require eight days. When all cells are occupied, the comb is removed from the comb cage and returned to the brood nest of the colony. A second empty comb and the queen are put into the cage and the same procedure is repeated after another six to eight days. On approximately day 18 the third comb is removed and the queen is returned to her colony. The mites emerging from the old combs in this colony will only infest brood cells open for colonization in the three trapping combs. As soon as all cells have been capped, or at the latest before the first bees hatch, these three combs are removed from the colony.

Similarly to the total brood removal method, the combs are either rendered or used for a split. In this case the new colony must be treated with oxalic or lactic acid which is less corrosive though not as efficient as oxalic acid in removing the many mites.

The original colony, from which the three trapping combs were removed, will be as strong as colonies that have been treated with formic acid at the beginning of the winter period. The loss of one bee generation will be balanced by either increased brood production and raising of young bees, and/or by prolongation of the bees' life span.

Caging the queen

The final method involves caging the queen for approximately 21 days. To do this, a hole is cut into a central brood comb, into which a special cage is inserted.

The queen is transferred to the cage, and thus prevented from laying eggs. The holes of the cage have the same size as in queen excluders. The bees can feed and groom their mother from either side. After 21 days, or 24, if there is drone brood, all bees and all mites will have emerged from the brood cells. The cage is taken out and the queen freed. The remaining hole will quickly be rebuilt again.

It is now imperative that the colony is sprayed with oxalic acid or lactic acid to remove all mites on the bees. Again, the missing generation of bees has no consequences for the colony strength at the beginning of the winter season.

The caged queen is prevented from laying eggs for three weeks. After this period all bees have hatched from brood . The colony is thus free from brood and can now successfully be treated against varroa.

Thoughts on the future

We are convinced that biological methods will prevail in the future, because they do not involve formic acid and consequent damage to the brood, and this is possibly also true for oxalic acid most of the time. There are two reasons why biological methods are stepping stones on the path towards Varroa tolerant and resistant colonies: firstly they help increase the vitality of the colonies by avoiding significant negative side effects of these acids. And secondly they will lead to a comparatively large mite load in the colonies during the period when the winter bees are being raised between August and the end of October. The presence of the mite is an important precondition for the evolution of tolerance and resistance. Both are stepping stones on the path, which we beekeepers want to take together with our bees.

American and European foulbrood

American foulbrood time and again infects colonies in Germany and it is classified as a notifiable brood disease. Its visible symptoms are gaps in the brood nest and sunken caps on brood cells. The so called 'matchstick test' reveals a sticky mass in infected cells, which can be drawn out into a thread. A serious infection results in an unpleasant smell. If a case is suspected, it is necessary to contact the responsible vet or beekeeping authority, who take charge of further steps to deal with the outbreak. Infected colonies and, depending on the opinion of the authorities, all neighbouring asymptomatic colonies have to be salvaged by means of the artificial swarm method, or they have to be destroyed. The affected apiary is shut down, and an standstill area with a radius of at least one kilometre is imposed.

European foulbrood, also known as sourbrood, is also a notifiable disease in Switzerland, but not in Germany. It also shows gaps in the brood pattern, and in more serious infections produces a typical sour smell. Larvae that have reached the curled stage in open cells change colour to a dull yellow, which later turns to dark brown and they become twisted and dried out.

Both diseases are a catastrophe for the beekeeper affected, as well as for the colleagues in the standstill zone. For the beekeeper, because all his colonies will be destroyed if they are badly affected. For the colleagues, because moving colonies into the area or out of it is prohibited. Quite frequently tensions run high between beekeepers in the affected region even though those whose hives were diagnosed with the disease in most instances could not have done anything to prevent infection.

It would help to lower such tensions if the regulations were to be changed. Colonies which show only limited impacts of foulbrood can be salvaged with the artificial swarm method. This involves confining the bees to a basement for three days, so that the honey containing the spores of the disease shifts from the stomach to the digestive tract of the bees. The pathogen is then disposed of from the rectum outside the hive. The artificial swarm is put onto empty frames, where the bees will 'sweat' themselves healthy in their urge to build comb.

The Golden Hive and teaching beekeeping

Honeybees are a valuable aid in education. They help children and teenagers to understand the task and role of all living beings in the context of nature. Karl von Frisch, one of the most prestigious researchers into the behaviour of bees, writes in his book 'On the Life of Bees': "Whoever can maintain an open mind for nature in the midst of rampant technology will experience an insight into the life of bees as a source of joy and amazement. For the beekeeper this will be the basis for success. For the teacher who wants to implant a love for the living world into young minds, it will provide the most beautiful subject matter." Why is this the case? Here the uniqueness of honey bees may be emphasised by a few facts which are worth following up in a pedagogical context.

A bee colony can be seen as a unit only in thought. Observations reveal three separate bee organisms, the workers, the drones, and the queen. But only when seen as a whole colony are they linked functionally and socially with each other. From its disposition a bee colony is immortal. The three individual beings are renewed at different time intervals in order to secure the colony's survival.

The threats to the survival of insects and bees, some of which have global reach, are man made. Beekeepers, who keep colonies for the purpose of exploitation and maximising honey production, are implicated.

These few aspects show the various ways in which a love for nature can be kindled among children and teenagers, and how it is possible to enable them to learn from nature.

In this chapter we want to discuss the characteristics of the Golden Hive in an educational context that supports a phenomenological approach. In other words, we are not dealing here with general beekeeping practices that have been outlined in other chapters, but we want to present their advantages when working with children and teenagers.

The Golden Hive is particularly well suited for when experiencing the a bee colony is the main focus. Many such hives are already located next to schools and even Kindergartens, where it is possible to watch from the outside the bees returning from foraging. The hive will be opened from time to time under teaching supervision to allow light into the darkness of the bees' home. It is thus possible to see how thousands of bees live and work together in their shared home. The act of opening the hive should already be something special. Perhaps it is possible to begin with a small ritual such as a gentle knocking or reciting together a bee related verse or poem to indicate that a bee colony deserves respect. Children have an understanding for this. They too, are reluctant to show their rooms or the home they live in to strangers whom they don't know.

Bees prefer calm, steady movements when lifting the frames, and a visit to the colony should not last too long. It is not possible to provide universal guidelines, because individual colonies differ in their response, and they also vary in their temperament. Experienced beekeeping teachers will recognize the moment when the bees become agitated by their climbing up from the spaces between the combs to the tops of the frames. This situation can be explained to the children prior to the opening of the hive, because they too know very well that they have different moods, depending on whether they got out of bed on the wrong foot or not. So if the group opens the colony with the question in mind "what mood are the bees in today", all attention is focused on this issue and the children themselves gain more and more confidence. In time they can even notice that the bees react differently to the person, who is dealing with them. Some children and the bees can approach each other with ease, whereas others don't so easily interact. Here coercion should be avoided. Young observers, who prefer to stand at some distance, can still see a lot and gain an overall impression. If the sun is out and the weather is fine for foraging, and if the hive stands by itself, it can remain open for up to ten minutes or even a bit longer. Experienced beekeeping teachers have a sense for what is right and follow this when they are opening a Golden Hive.

Setting up a Golden Hive

Bees can be placed at the edge of the school grounds or beside the playground of a Kindergarten. The direction of the entry hole should be arranged in such a way that the bees fly away from the grounds, if possible to the east, or the south east. In order to allow for the observation of the activity at the entrance hole, space should be available on both sides of the hive so children can stand and watch. Watching the hive entrance will already result in many interesting questions, e.g. about the varying colours of the pollen loads brought home by the bees on different days, or sometimes even at different times of the same day.

When siting the hive, a location in the blazing midday summer sun should be avoided. Otherwise the bees are forced to collect a lot of water to cool down the hive, and this may result in stress and thereby reduced docility in the interactions with humans. When located underneath a tree with sufficient shade, the morning sun will still reach the entrance hole and entice the bees into the open for foraging.

It is advisable to place the Golden Hive onto a stand made of wood or stones, taking care that the height allows the children, depending on their age, to see inside without having to lift frames.

As soon as the hive is opened, the first bees will come up and look the children in the eye and vice versa. Even though insects differ a lot from mammals, children nevertheless find them of great interest. By contrast with other hive systems, as there are no additional boxes or supers on top of the Golden Hive, the hive box remains the same size, and becomes a familiar fixture in the grounds of the school or the kindergarten.

In preparing for the arrival of the bees, the children can colour the hive together with wax crayons. This decoration makes it attractive for the children and for other visitors or friends. The colours also help the bees in their orientation and recognition of their own home, because they can see colour and distinguish patterns.

When siting beehives at schools or at kindergartens it may happen that a child gets stung. Although allergic reactions are rare, it is necessary to check with the parents whether an allergy is known, so that you can react quickly in an emergency. People who teach about bees have various ways of dealing with stings from simple consolation, to dabbing the stung area with an onion that has been cut open, or with a chewed leaf of ribwort plantain, or with Apis Globuli or a specific bee allergy medicine.

A visit by school pupils to a beekeeper is regulated by the rules of schools in the same way as visits to any other craftsperson. A similar situation applies to kindergartens. Issues in this context are duty of supervision etc., which will be known to the teaching staff. It is important to explain the educational value of bee studies at a parent teacher meeting prior to a visit to an apiary or to publicising in the classroom the offer of a beekeeping working group. It is important to make explicit that getting stung is a real possibility in exceptional cases. Any reservations the parents have must be taken seriously, but at the same time teachers should demonstrate their confidence that they are capable of dealing competently with such a problem. A beekeeper will know which of his colonies has a tendency to sting and will keep those hives shut. Colonies which are opened may sting, but they will only do so if they feel understandably disturbed. Sometimes it may be a good experience that those who tease or even molest bees in front of the hive entrance will get to feel a sting.

By decorating the Golden Hive, children can prepare for an imminent arrival of the bees. The colours and shapes will later on help the bees to orient.

A local swelling after a sting is far from being an allergy. Only nausea, vomiting, itching, and even breathing problems are indications of an allergic reaction. Most people who are affected like this are aware of their condition and will carry medication which can help immediately in such an emergency until medical personnel arrive to take over. No child is exposed to danger to such an extent that it would justify preventing them experiencing bees in educational settings.

Opening of the hive

Because opening a hive makes its contents immediately visible, the various layers on top of the frames and their functions should be explained during the process of opening. The cover of the Golden Hive consists of three layers. On the outside rain, snow, and wind are kept away by the weatherproof lid. Below this is located the soft fibre particle board for insulation, and below this is a cloth made from beeswax. Beeswax is also used in houses as a breathable ecological protector of wooden floors and furniture. Some of the children and teenagers may be familiar with this. From the perspective of the bees, the wax cloth is the top of their living space, i.e. the ceiling. Before it is peeled off, the pupils may put the back of their hands onto it. The temperature encountered there is frequently higher than that of our hands, and it can feel comfortably warm. This is not an expected sensation when dealing with insects.

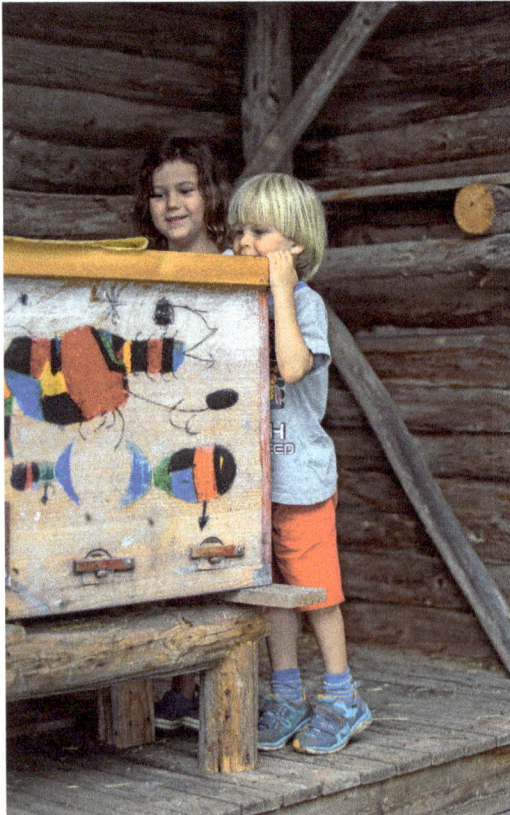

Also prior to the first visual inspection comes the experience of the fragrance of the hive as the wax cloth is slowly removed. Even though children tend primarily to visual experiences, and want to see the bees straight away, we should make sure that the aroma from the hive does not pass unnoticed.

Around midsummer, when the colony has the largest number of worker bees and drones, there is room in the Golden Hive for all frames next to each other. Closest to the entrance is the brood nest, and away from it are the honey stores. Because of this arrangement, it is easy to open only certain segments of the Golden Hive and to lift only selected frames. The other parts stay covered by the wax cloth, so the bees can continue doing their tasks without being disturbed. As a consequence, the colony overall will remain calm, even in the section being opened.

Apart from many bees, in the brood area we can sometimes see the queen. There are always eggs, small floating in royal jelly, or even fully grown fat larvae. In other areas, under permeable waxen caps, pupae are metamorphosing into fully grown worker bees or drones. Often hatching can be observed – a young bee chews her way out through the waxen cap and climbs onto the comb surface where she is immediately welcomed by her sisters. She quickly joins the throng and we lose sight of her. Located above the brood area, in the shape of a half moon, there is the colourful storage area of pollen. The more colours are represented, the broader is the range of forage from which the pollen has been collected. Just as we humans speak gratefully about our daily bread, so also beekeepers refer to the colourful pollen stored in the cells as 'bee bread'. Higher up, towards the top-bar of the frame is a small strip of stored honey that has already been capped by the bees. The ripened honey in the closed cells will last not only through summer, but also through the winter months.

By contrast to the waxen caps over the brood cells, the lids on the honey stores are not permeable to air or water. The differences in the structure of the caps can easily be observed. The band of honey on top of the brood area helps prevent the bee colony from starving even in periods of bad weather, or in a very dry summer when no nectar is available.

We can now close this part of the hive again with the wax cloth, and instead open up the area where the honey is stored. This allows the nurse bees in the brood area to continue their tasks of looking after the brood, feeding and warming it without further disturbance. We now carefully lift the wax cloth from the part of the hive furthers away from the entrance. The heaviest combs of a colony are found in the honey storage area, if they are filled to a large degree with capped honey. Once the children have had the chance to feel the weight of these combs in their own hands, they are deeply impressed by the diligence of the little bees who have collected all that honey, particularly when they are told that each bee can carry at most 50 milligrams of nectar per forage flight. The combs with honey are mostly free from bees, as there is no work for them there in summer.

Honey fresh from the comb is a special delicacy: it is squeezed out and licked off a finger pressing across the capped cells. It is also noteworthy that honey from different combs from the same colony can differ in taste. Why this may be the case is another topic that can be discussed with the children.

Harvesting the honey

Honey is the ultimate bee product par excellence, and it is fun to harvest it with the children. Depending on the location of the hive, the bees may collect so much honey that it is obvious that they have more in the hive than they need for the winter period. In this case honey combs may be removed and eventually replaced with empty frames containing foundation or empty combs, in order to give space for nectar still to come. If the bees are allowed to over-winter on all of their collected stores, we can remove and extract honey from any superfluous honey combs in spring, when the flowers are again providing sufficient nectar. It is appropriate to discuss with the children, even from an ethical perspective, the topic of honey harvesting and the amount required for overwintering, so that they get a feeling of what it means for bees when vital stores that they have collected at great effort are withdrawn from them.

The transportation to the extractor is facilitated by a light-weight comb transport box , which can easily be moved with a wheelbarrow. More on the harvesting of honey on page 147.

Extracting honey together with children

Uncapping and centrifuging honey is a special experience for children of all ages. Once they have received a proper instruction, they can uncap honeycomb themselves with an uncapping fork. They will understand that cappings wax is the most pristine wax because it has been freshly sweated and contains no contaminants. The children can collect this wax wet with honey and put it back into the hive, where the bees will rapidly clean and dry it and return the honey in it to the colony.

They will also eagerly learn how to operate a manual extractor. In order to avoid comb breakage, the initial speed will be low. The combs are then turned round, and a second slow rotation is applied. Once the bulk of the honey has come splashing out, the combs can be spun faster and faster (see pictures on pages 100 and 101).

In between, the 'liquid gold' is poured through a double sieve into buckets. Now is the time to taste the freshly centrifuged honey, and with the older children to measure its moisture.

Managing and feeding of the bee colony

As we have seen, it is easy to expand bee colonies in Golden Hives simply by adding single frames. It is stimulating to discuss in advance with the children the reasons for such an expansion and estimate together, based on colony development, whether the bees may need even more space in the following weeks. Estimating correctly is a challenge, which will often only become obvious at the inspection.

If too much honey has been taken from a colony, or if insufficient stores have been collected for a successful overwintering, it is necessary to feed sugar syrup. In a Golden Hive, feeding is simple and can easily be managed by children and teenagers. Mixing syrup is a task that even small children can handle (for the recipe for syrup for feeding see page. 74).

A small bucket with some straw is put behind the dummy board into the space that has become available when the size of the colony has been reduced to 10 or 12 frames. The straw prevents the bees from drowning. The bucket can be filled almost to the brim with the syrup. The bees find the bucket and take the food to the combs in their colony by crossing underneath the dummy board.

When feeding, the Golden Hive should be opened only for a very short period in order to prevent bees from other colonies from entering and robbing. If the children are too young, placing a feeding bucket may only be done by an adult. The same applies to carrying heavy loads of feed. If feed is spilled onto the ground next to the hives, it encourages robbing. It is possible to discuss robbing with the children from grade four onwards. An experienced bee veterinarian commented on robbing as follows: "In late summer, the bee colonies test each other for vitality." In other words: this behaviour is part of the natural selective criteria that lead to healthy bee colonies, provided it was not provoked by basic mistakes like spilling sugar solution or opening of a hive for too long.

Varroa treatment of colonies in school grounds

Treating bee colonies with organic acids in school and kindergarten grounds is difficult for all involved because of the emotional connection of the children with the bees. It is nevertheless necessary to treat in order to avoid as much as possible the death of the colonies. If children or teenagers younger than grade ten age are involved, it is better to let adults treat the colonies in their absence . Dead colonies should also be cleared away by adults. Please note that formic and oxalic acids are quite dangerous, and remember that advised safety measures have been designed and formulated for adults, not children. It is important to store the acids safely and out of reach of children. The pros and cons of varroa treatments can be discussed with the older grades. They are able to understand the need to keep the colonies alive, even if the bees clearly show that the treatment is an imposition. Treatment can easily be done in late summer with a Nassenheider evaporator fixed into a frame (see page 120). The same precautions apply to all winter treatments with oxalic acid.

In summary we hope that the simple beekeeping methods associated with the Golden Hive will convince many educational professionals that children and teenagers can experience bees without problems.

∾ Summary of the advantages for bee-related education
The Golden Hive is ideally suited to encouraging joy and understanding for bees among children and teenagers:
▶ The wax cloth cover allows the sensual experience of warmth of a brood nest.
▶ When pulling back the wax cloth, the fragrance of the hive emerges – a smell sensation for all.
▶ Approach to the brood nest proceeds slowly, the bees are hardly stressed.
▶ Harvesting of honey combs takes place without disturbing the colony.

Products from the bee colony

Because they are masters of transformation, to this day honeybees are received with great sympathy. They turn the easily perishable nectar that they collect with proverbial diligence into something new: sweet honey that keeps indefinitely. Elements of transformation can be found in all other products from the hive. The resin from tree buds is turned into propolis by bees; pollen from blossoms is transformed into bee bread. And beeswax, bee venom, and royal jelly are secretions, which are produced entirely within the bodies of the bees. The qualities of transformation can be appreciated in all products from the hive.

In addition, they all exhibit healing properties and they are still to some extent used as such, either directly or blended with other ingredients in medicines and cosmetics. Apitherapy, in which products from the hive and even its air are used for medical or therapeutic purposes, is an important part of complementary medicine. This is true even today, despite the fact that these products are frequently contaminated with chemicals from industrial agriculture, from the environment in general, and also as a result of beekeeper practices such as chemical treatments against varroa.

In order to have hive products which are residue free to the greatest possible extent, we have to pay attention of the environment in which our colonies live and forage, and which compounds or biological measures are applied to control the varroa mites. Many amateur beekeepers are very closely monitoring these circumstances.

With their larger number of bee colonies, and larger volumes of products from their hives, many professional beekeepers also feel obliged to follow criteria of animal welfare and are careful to sell products free from residues, even joining organic certification associations such as Demeter. From the association whose logos are printed on the packing, consumers can find information on the internet about the standards. With bee products bearing the label of an organic certification association, you can be sure that the producer is regularly audited, with regard not only to his or her treatment of the bee colonies, but also to the products offered which are assessed to provide the best possible food safety.

The Golden Hive has not been developed with the intention of giving the highest yields of honey or other bee products. As with the production of food items in general, it is important to achieve a fair and reasonable compromise when it comes to 'harvest' time. With this type of hive, a good compromise between animal welfare and the harvest of products is assured. One can definitely harvest sufficient honey to supply one's own family and friends. It is also possible to work wholly extensively with it and thus carefully harvest the honey only in spring, when the bees are able to get fresh supplies of nectar and pollen and hence do no longer depend on their winter stores. They can be removed without any negative impacts on the colony.

Delicious honey

On average people in Germany consume more than one kilogram of honey per person per year, about half the amount consumed in the UK. Only 30% of consumption in Germany is supplied by local beekeepers whereas in the UK it is as little as 6%. The far larger share of consumption is imported, mainly from eastern Europe, Mexico and South America.

Bees collect honey for their food requirements. A colony consumes it in large quantities in order to be able to engage in all necessary activities, including heating and feeding the brood, building comb, and foraging. According to scientific estimates, beekeepers never get to see 70% of the total amount of nectar collected, because it is immediately metabolized by the bees. So all the more valuable is the honey stored in combs. It represents the winter food from December to the beginning of March when little or no forage can be collected. If we remove that honey before the winter period, we have to provide a substitute without delay.

Time and again beekeepers who want to overwinter their colonies on their own honey stores ask whether it is possible to keep bees without feeding. Our answer is brief and decisive: "Yes, this has worked well most of the time for the last 50 million years." However, it is important to take into account that nowadays in some regions where there is no forage available in summer and especially late summer, it is not certain that bees can collect sufficient stores for winter. In such instances it is necessary to provide supplementary feed, even if no honey is taken from the hive.

We would like point out again that, in a natural setting, if the first honey in spring is stored in capped cells, it serves as stores for the winter. It is followed by honey from summer forage, and in some regions followed by honey dew. Because of the large combs in a Golden Hive, harvesting monofloral honeys is hardly possible – e.g. rape or cornflower honey – which, according to the honey directive, has to be sourced completely or predominantly from the specified flowers or plants. Such honeys have to show corresponding organoleptic, physical-chemical, and microscopic characteristics. This is usually impossible with large combs, which under ideal conditions can store three kilograms of honey.

Because of the fact that honeydew puts a strain on the rectums of the bees due to its high content of minerals, it is not well suited for winter storage. For this reason, the Swiss bee research institute recommends restricting the amount of honeydew winter stores to 30 per cent. Beekeepers who want to overwinter their colonies on their own honey, should remove capped combs with floral honey, and swap later combs with honeydew. Combs with honey stores must – prior to reintroduction – be kept in a cool, dry, and dark place. Putting them in a freezer works, too.

It is possible to harvest honey without an extractor. Natural combs with honey can easily be cut into pieces to be consumed as comb honey. Wax and honey separate from each other during chewing. This is a beautiful method since up to the point of consumption the honey remains in the original packing provided by the bees. The wax may be spat out after having chewed the honey.

It is also possible to cut honey comb from the frames with a knife, cut it up into chunks and stir the mass with a large wooden spoon. The honey is then allowed to drain overnight into a bucket or pot through a fine-meshed straining cloth obtained from beekeeping supply companies. The remaining wax can be placed into the hive in a bucket, so that the bees can lick off any remnants of honey. The clean wax is then melted. Honey from this draining method contains a larger amount of pollen and results in a very special character.

Honey extracted by a centrifuge is first poured through a double strainer. Fine wax remnants are removed, the honey drips into a honey bucket.

ॐ The harvesting of honey

In order to harvest honey, the combs from a Golden Hive can be extracted with a centrifuge for Dadant sized comb available from most beekeeping supply companies. It is important that the combs fit into the comb basket. Despite the presence of mushroom shaped frame spacers, the combs should rest completely against the wire basket, otherwise there is a danger that they will become distorted or even break. For this reason, for the fragile naturally built combs, it is that a tangential centrifuge rather than a radial one is used.In the latter full centrifugal forces of the whole comb bear on its edge, whereas in the former case the force is distributed over the whole area of the comb. In order to be on the safe side regarding the size of the wire basket, it is recommended to take a frame along when purchasing a centrifuge.

The large area of comb per frame requires some stabilization, e.g. by means of stainless steel wire embedded in the comb. Frames with wiring, particularly if they have already contained brood once before, are more robust when being extracted compared to freshly constructed virgin comb. It is necessary to gradually adjust the speed of rotation in order to prevent breakage of the combs by the centrifugal forces. Slow rotation and turning the combs round in the wire cage several times is beneficial for the preventing of comb breakage.

After centrifuging, the honey is poured into a bucket via a double sieve, coarse and fine mesh. It should be made of stainless steel or food grade plastic (more on this on pages 100 and 101).

In order to achieve a very creamy product, we remove the foam from the top of the honey with a large spoon or a food spatula while it still is liquid and transparent. Beekeepers refer to this process as skimming.

As soon as the honey starts to crystallize, stirring begins, which results in a somewhat milky consistency. Stirring takes place over a period of five days, three to five times every day, for 10 minutes. Various tools for manual or electrical stirring are available. However note that some honeys solidify very quickly, and it may then be impossible to fill them from the bucket into jars. In other words, filling the jars needs to be done before the honey becomes fully crystallized.

Beeswax and combs

Beeswax is a natural product sweated out by the bees from glands on the underside of their abdomens. It is malleable, soft, and hardens quickly into small scales. These scales are clear and transparent with a white border, similar to quartz crystals. A magnifying glass helps in seeing the wax scales on the floor of the hive, after they have fallen there in the process of comb construction. Hundreds of bees are required for the process and act as a 'comb construction gang', producing beautiful, white, virgin combs. It is only by their use as a nursery, store, or dancing floor that the comb turns yellow, brown, and eventually almost black. The bee researcher Jürgen Tautz has summarised very fittingly the role of comb in his book 'The Buzz about Bees": "Properties of the comb are integral components of the superorganism that contribute to the sociophysiology of the bee colony.." The combs of a colony carry the whole range of social interactions of the bees for years to come.

Apart from oil lamps and open fires, for hundreds of years candles made from beeswax were the only source of artificial lighting. Large numbers of them were required, especially in ecclesiastical and religious contexts.

Nowadays companies producing cosmetics and pharmaceutical items are the most important buyers of beeswax. They turn it into creams, ointments, lotions, and lip balm, or use it for heat wraps and soothing salves, which also contain other components from the hive such as propolis and pollen. Beeswax is very well tolerated by human skin, making it supple, and providing external protection. Once the bees no longer need the combs, they can be harvested and rendered without further ado.

✿ Melting beeswax

Following the honey harvest the wax residues can also be put to use. The easiest method is to put the frames with comb, pieces of wax or wax cappings from honeycombs into a solar wax extractor where it melts under a glass panel. It flows through a grid to clean it from debris and impurities, and is collected in a stainless steel container. All commonly available solar wax extractors are big enough to hold at least two to three Golden Hive frames. As with buying a honey extractor, you are advised to take a frame with you to the beekeeping supply store. The wax collected in this way is liquefied again in a water bath and further purified by pouring it through a cloth. The homogenous product is suited for the production of salves and beeswax candles. It can also serve to make foundation, starter strips, or in making wax cloths for covering the top-bars. There are a number of manufacturers offering to process even small quantities of wax.

After a one-time usage for foundation, it is advisable to exclude it from recycling for the hive. Pesticides applied in agriculture as well as chemicals used against varroa accumulate in beeswax. This may be problematic both for medical and cosmetic purposes, and also even when making candles.

Precious pollen

Research by the bee institute in Liebefeld, Switzerland, shows that only bees which are well supplied with pollen can raise healthy, long-living sisters. Nurse bees must therefore have sufficient and diverse pollen available in order to produce nourishing brood food and royal jelly for the brood and the queen. Pollen contains compounds with antibiotic effects, proteins, carbohydrates, fats of lipids, vitamins, and mineral trace elements. A variety of these ingredients protect both bees and humans against diseases. In other words, pollen is a highly valuable natural product. Pollen can help people desensitise when they suffer from pollen allergies. This is best achieved using one from local flora. Harvested pollen must be properly cleaned and dried. Prior to consumption it is soaked in water, milk, or yoghurt for a number of hours, otherwise it is almost impossible to digest.

Nutritious bee bread: The variety of colours show that the bees have good and diverse stores of pollen in their cells.

❧ Harvesting pollen

With care, pollen may be harvested by various methods. The most common is the pollen trap, a device that is hung in front of the entrance of the hive. It forces the bees to crawl through holes in a grid on their way inside. The diameter of the holes is such, that the pollen loads are stripped off the forager bees and then fall through a mesh into a drawer below.

A modest supply of pollen remains available for the bees, because some of them always manage to crawl through the holes without losing their package. The trap is emptied on a daily basis and is left in place for at most three days. The colony then recovers and the procedure may be repeated again a few days later.

A pollen picker enables stored pollen, so-called bee bread, to be collected from cells in the comb where it has been stored. This is a tedious procedure and requires patience. It only makes sense for the needs of the beekeeper and his or her family, because economically it would never pay, even though the pollen is worth having. By contrast to pollen from a trap, bee bread has been processed and thus can be eaten without soaking.

A professional method is to shred deep-frozen combs containing pollen. In this process wax and pollen are separated in a kind of centrifuge immediately after they have been shredded into small pieces. This method is used by some of the bigger apiaries for harvesting on a commercial scale bee bread which often occurs in large quantities. It would be a pity to simply render the pollen together with the combs. After processing bee bread is sold under the brand name 'Perga' especially in health food shops.

Propolis for bees and people

Only a few bees of a colony collect the sticky resin from trees and shrubs. They transport it in a way similar to that by which pollen is brought home as loads on their hind legs. Inside the hive, the resin is chewed and kneaded by the bees with their mandibles, and in the process the resin is blended with a special salivary secretion, and turned into propolis. So far more than 200 components have been identified in it, and it is likely that more will be found.

There is no pharmaceutical licence for propolis. It should not be advertised as a remedy for particular conditions. The chemical composition of propolis varies between different species of deciduous trees, and also from year to year within the same tree species. For the lawyers of the European pharmaceutical legislation this presents a nightmare scenario, because they would want to approve products consisting only of exactly defined ingredients present in exact quantities.

The bees cover the area surrounding the hive entrance hole with. The result is a disinfecting doormat which prevents pathogens from being brought into the hive by returning bees. A thin propolis coating is also found on all hive surfaces on which the bees walk – on the walls, the frames, and on the edges of the combs.

Propolis is considered to be one of the most effective, natural antibiotic substances. Not only does it kill bacteria, but it also inactivates fungi and viruses. Therefore this product from the hive is also a natural remedy. In most instances it is offered for sale as an alcohol based tincture or as a salve, and it can be purchased in pharmacies as a food supplement. Beekeepers are allowed to harvest propolis, but under no circumstance must it be sold as a medical remedy. If using it, it is also important to know that some people show allergic reactions to some of its ingredients. This risk should be checked for prior to a first application.

❧ Harvesting propolis

For small amounts, harvesting propolis is possible by scraping it from the top bars of the frames. The bees tend to place it in this location mostly from the middle of summer onwards, in order to close gaps between the frames and the wax cloth on top. Amateur beekeepers therefore harvest propolis during their inspections from spring until the end of July. Later harvests are problematic, because towards the approaching winter season all gaps, cracks, and slits need to be closed with propolis to reduce any cold draughts.

Larger quantities of propolis are harvested by means of a special propolis screen made from food grade plastic, which is placed underneath the wax cloth on top of the frames. Once the bees have filled all the gaps in the screen, it can be removed and put in a freezer. Once frozen, the pieces of propolis pop out easily when the screen is twisted and bent. This is called 'raw propolis', because it also contains remnants of beeswax.

For one's own use it is possible to extract the propolis from the wax in 70 per cent solution of alcohol. Any suitable amount of raw propolis, crushed for optimal extraction, is put in a bottle and mixed with an equal volume of alcohol. The capped bottle is shaken daily for three weeks to extract the propolis. The mixture is then filtered through a coffee filter to eliminate wax and other insoluble particles. The resulting tincture is transferred and stored in small, preferably brown bottles with drippers.

We have done trials over a number of years with various propolis screens. We have found that it is possible to harvest 80 grams per colony in a Golden Hive per year without recording any stress symptoms in the colonies.

Advanced beekeeping

A few years ago during a meeting with pioneer users of the Golden Hive, the purpose was to exchange ideas about the improvement of the hive. Among the accumulated beekeeping experience in the meeting there were a few voices who really wanted to harvest the arch of honey at the tops of the brood frames. As we have seen, unlike in the more common hive systems, in the Golden Hive brood and honey can be found on the same combs. In order to provide an alternative for beekeepers who want a clear separation between frames containing brood and those containing just honey, it was suggested that two honey supers be placed on the Golden Hive. This would allow the bees to store honey in the supers, and care for the brood below them. The size of a honey Golden Hive super frame has the height of one third of a brood frame, i.e. 150 mm, with the width kept at 285 mm. A few colleagues immediately implemented this suggestion on returning home. Positive feedback came during the nectar flow in spring: the honey is indeed stored in the honey frames on top, while the brood nest is built on the large frames below.

With this change, the principle of one inclusive hive body was given up in favour of a hive divided into three parts. Depending on the strength of the nectar flow it is possible to make only one of the two honey supers accessible to the bees. For that, the wax cloth is cut into two equal halves. One half is placed on the top of the one honey super, the other remains under the second, which keeps the bees from entering it.

However, as soon as harvesting honey is given priority, many of the advantages of the simple way of keeping bees outlined in this book no longer apply. When checking for signs of swarming, or for any other brood chamber inspection, it now becomes necessary to remove the two honey supers, and then to replace them afterwards, as is the case with Dadant or Langstroth hives. In order to support the softwood particle board and the metal cover on top, it is essential to have both honey supers next to each other on top of the hive. Years of experience have shown that 95 per cent of all beekeepers who work with the Golden Hive are happy without using honey supers, because they appreciate the advantages of an extensive method of beekeeping without the

A single cavity hive is turned into a three cavity hive. The honey supers are always placed on top in pairs. Depending on the nectar flow, honey often is only harvested from the super which is located above the brood nest. The second honey super is kept in place simply to retain the fit for the metal roof and the insulation board.

need for shifting supers back and forth.

We did not want to omit mentioning the option to separate the honey and brood areas more distinctively. It is compromise solution for 'honey hunters'. Obviously, any such solution needs to retain the same frame size in order to be compatible with the original Golden Hive, e.g. to enable the exchange of split colonies with other beekeeping colleagues.

Ten-frame-hive for young colonies

At the apiary of Mellifera, there were times when we were managing more than 100 Golden Hives. In the swarming season, in order to cope with the many young colonies resulting from prime swarms and queen splits, we developed a ten frame hive. A split containing two frames with bees and brood, along with two empty frames, plus a dummy board, i.e. a total of five frames, is much easier to carry than the full-sized Golden Hive.

Once removed to the apiary reserved for young colonies, a frame feeder is added. It has the same dimensions as the standard brood frame, but holds a square plastic for liquid feed. The advantage of this system lies in the fact that, placed next to the brood frames, it reduces the distance that the bees have to travel to access the food. Under ideal circumstances these colonies reach a size of eight to ten fully built combs by the end of summer, and can overwinter in the ten combs box.

The honey supers described on page 157 fit onto these ten frame boxes. This makes it possible to use such a hive with one super placed on top during the cherry blossom period in spring for harvesting a varietal honey. However, this method is even further away from the original idea of the Golden hive, namely that it is a horizontal hive. In effect it is managed like any standard frame system, the only difference being that the frames in the brood area are oriented in portrait format. Due to the very small frame size in the honey super it is possible to place any number of supers on top of each other. This facilitates the harvest of monofloral honeys, mainly because the honey in the small frames will be quickly processed and capped by the bees.

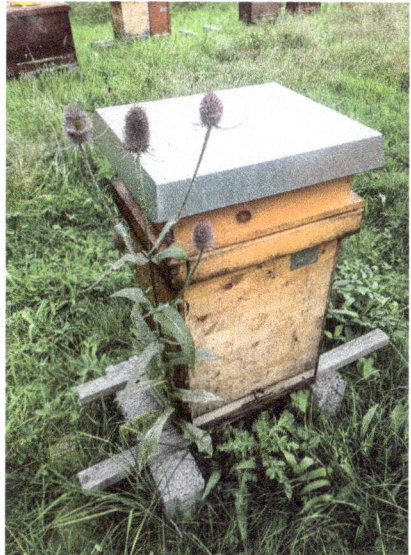

The small honey super as a mating nuc

No colony will ever make use of all the 15 to 20 queen cells produced in the run up to swarming. Experienced beekeepers may wish to produce more young colonies than those made from queen cell splits. This enables young queens to be kept in case of an unexpected and sudden loss of a queen in a full sized colony. That way, interruption of the breeding process is kept to a minimum, because the large colony does not have to raise an emergency queen. Sudden loss can happen when the queen is injured or squashed during an inspection.

The small honey frames are ideally suited for the establishment such mini colonies. Two empty frames plus a small feeding frame can be put into a honey super which is placed onto a purpose-built hive floor. The size of the frame with the feeder acts as a division board, i.e. it must fit tightly to the walls and the top. Two empty frames with starter strips are placed next to the entrance.

A queen cell close to hatching is cut from a brood comb, leaving plenty of margin, and it is then firmly fixed to the frame which is further back in the nuc. By adding a quantity of bees that would fit into a soup ladle, this tiny colony can take care of the queen until her mating flight. A small brood nest will appear on the two frames which by now will have been completely filled with comb. The feeder should not be allowed to empty, because the colony always needs to be kept well supplied. A few rainy days without forage flight would threaten its survival.

With sufficient feed the nuc will grow quickly, and step by step the feeder must be pushed back and empty frames need to be added in front of it. Once the colony has filled nine frames, the feeder is removed and replaced with a honey super containing 10 frames. From now on feeding is only possible from the top with the feeder placed in a second empty honey super. It is possible to successfully overwinter such small colonies.

In a small honey super with two frames, a generous hand full of bees can well look after a queen that has not yet emerged. The feeding pocket, which is placed at the back end, has to be well stocked with fondant. There are sufficient stores if bees are prevented from foraging for a number of days.

This small frame from a honey super of a Golden Hive was turned into a mating unit. The frames with starter strip and emerging comb are the basis for the growth of the small bee colony.

Learning from the bees

The key constants in the Golden Hive are the uniform frame dimensions and the number of frames next to each other. As time went by the saw a number of modifications. They were result primarily of our own observations, and close observations by others of the behaviour of the bees in the hive. Not all suggestions have been implemented. There are various reasons for this. For one, the hive had to remain easy to manage, highly functional, and practicable. Its cost-benefit ratio convince the largest possible number beekeepers to use the hive. Specific modifications such as a vertical queen excluder do not accord with our idea of bee appropriate beekeeping. But they can still be implemented by individuals through simple adaptations. And in any case, we hope that beekeepers will continue to tell us about their experiences and share their suggestions. We also look forward to seeing excellent photos and/or reports on beekeeping in Mellifera's Golden Hive.

A number of initiatives, discussion groups, and seminars have arisen in connection with the Golden Hive, all originally originating at the research and teaching apiary at Fischermühle.

The network 'Bienen machen Schule' ('Bees as teachers') brings together educators who want to learn beekeeping to teach it in their schools, and beekeepers who would like to share their skill with teachers. The network 'Blühende Landschaft' ('Flowering Landscapes') has been working since its founding to improve the availability of forage for bees and other pollinators. It engages with citizens, communal authorities, farmers and gardeners aiming to share knowledge and encourage action.

Regional Mellifera groups exist all over Germany, where beekeeping experience is exchanged and the magic of bee-appropriate beekeeping shared on a regular basis. This is also where activities are organized such as the Berlin conference 'Learning from the Bees', or participation in demonstrations against the increasing industrialization of agriculture.

In the Mellifera Training Association colleagues have come together to working on teaching and ongoing training of beekeepers. Despite the wide individual leeway in the types of training, all beekeepers involved wish to popularise the origins of bee-appropriate beekeeping and to contribute to its ongoing evolution.

After all, all of us in the large Mellifera network have probably been enchanted by the bees. We are always happy in this world of bees to dive in and be given unimagined energy and joy. One participant in a course has put this as follows: "The magic of the bees has taught me to take a fresh look at myself, the circle of my friends, the environment surrounding me and agriculture. Bees connect it all."

It is our hope that with our book we succeed in generating enthusiasm among more and more beekeepers for our way and our philosophy of beekeeping. We are confident that carefully handling the colonies will leave its imprint far beyond the confines of beekeeping.

The authors

Master beekeeper[5] **Norbert Poeplau** is responsible for managing some 300 bee colonies at Mellifera Association's teaching and research apiary at Fischermühle, Rosenfeld, Germany. His comment: "The Golden Hive is by far my most preferred hive type for working with bees.".«

Dr. Johannes Wirz holds a PhD in molecular biology. He is the co-director of the Natural Science Department of the Goetheanum in Dornach, Switzerland. He is a passionate beekeeper and co-ordinates the research activities of Mellifera Association in which he is also a trustee.

What started out as a teaching and research initiative for bee-appropriate beekeeping, has grown into an umbrella association in which a commercial apiary, the 'Network Flowering Landscapes' (Netzwerk Blühende Landschaft), and the educational initiative 'Bienen machen Schule' ('Bees as teachers') contribute to the success of the Mellifera Association.

Mellifera e.V.

---

5 In Germany, the term 'master beekeeper' indicates that a three year training program in apiaries and a college has been completed, plus a three year period of practical beekeeping. Master Beekeeper' is also a UK qualification: https://www.bbka.org.uk/education-and-training.

Acknowledgements

To the hundreds of people who have participated in our courses and seminars, we are grateful for their attention and questions that time and again have inspired us to re-evaluate issues and procedures, or even look for completely new approaches. A special thanks goes to our beekeeping friends Sabine Bergmann, Maja Högner, Johanna von Halem, Thomas Radetzki, Günther Mancke, Albert Muller, Markus Hilfenhaus, Uli Hampl, Martin Schäfer, Martin Dettli, Hugo Löffel, Erhard Maria Klein and Heinz Risse, all of whom tirelessly accompanied us and supported us in discussions in the ongoing development of the practical and 'philosophical' foundations of bee appropriate beekeeping. In a certain sense this Cupertino is a mirror image of the swarm intelligence and the wisdom of the hive that our bee colonies unreservedly share with us.

Appendix

German speaking readers may have a look on our website

Mellifera e. V.
'Biene Mensch Natur' Initiatives
Teaching and Research Apiary, Fischermühle
Fischermühle 7
72348 Rosenfeld
GERMANY
E-Mail: mail@mellifera.de
www.mellifera.de

🐝 **Other beekeepers using the Golden Hive (Einraumbeute) can be contacted through Mellifera Network:**
mellifera-netzwerk.de
The network has many nodes and anyone can see if there are Golden Hive users near them.

Further addresses

🐝 **The Natural Beekeeping Trust (UK)**
www.naturalbeekeepingtrust.org
Co-founder Heidi Herrmann

🐝 **David Heaf (UK)**
www.bee-friendly.co.uk
Modified Golden Hive: www.dheaf.plus.com/framebeekeeping/modified_einraumbeute.htm

🐝 **Jonathan Powell (UK)**
Tree beekeeper and Trustee of the Natural Beekeeping Trust
www.naturalbeekeepingtrust.org/product-page/the-tree-beekeeping-field-guide
www.freelivingbees.com
Email: jmapowell@gmail.com

Alex Tuchman (USA)
Director, Spikenard Farm Honeybee Sanctuary
www.spikenardfarm.org

Michael Thiele (USA)
www.apisarborea.com

Bees for Development (UK)
www.beesfordevelopment.org

Matt Somerville (UK)
www.beekindhives.uk
Produces modified 14-frame Golden Hives for sale: www.beekindhives.uk/the-golden-hive
Will make 22-frame version to order.

Leo Sharashkin (USA)
www.horizontalhive.com
Teaches natural beekeeping and publishes books on deep-frame hives: 'Keeping Bees in Horizontal Hives' by Georges de Layens and 'Keeping Bees with a Smile' by Fedor Lazutin.

Biodynamic Association UK
www.biodynamic.org.uk

Biodynamic Association USA
www.biodynamics.com

Further reading

Berrevoets, E. (2009) **Wisdom of the Bees: Principles for Biodynamic Beekeeping**. SteinerBooks

Bresette-Mills, J. (2016) **Sensitive beekeeping: practising vulnerability and nonviolence with your backyard beehive**. Lindisfarne.

Bush, M. (2012) **The Practical Beekeeper**. X-Star Publishing Company.

Freeman, J. (2016) **Song of Increase: Listening to the Wisdom of Honeybees for Kinder Beekeeping and a Better World**. Sounds True.

Frisch K.v. (1953) **The Dancing Bees: An Account of the Life and Senses of the Honey Bee**, Harvest Books New York, a translation of Aus dem Leben der Bienen, 5th revised edition, Springer Verlag.

Heaf, D. (2011) **The bee-friendly beekeeper**. Northern Bee Books.

Heaf, D. (2016) **Natural Beekeeping with the Warré Hive**. Northern Bee Books.

Heaf, D. (2021) **Treatment Free Beekeeping**. IBRA and Northern Bee Books.

Kornberger, H. (2019) **Global hive: what the bee crisis teaches us about building a sustainable world**. Floris Books.

Seeley, T.D. (1995) **The wisdom of the hive: the social physiology of honey bee colonies**. Harvard University Press.

Seeley, T.D. (2010) **Honeybee democracy**. Princeton University Press.

Seeley, T.D. (2016) **Following the wild bees: the craft and science of bee hunting**. Princeton University Press.

Seeley, T.D. (2019) **The lives of bees: the untold story of the honey bee in the wild**. Princeton University Press.

Siegel, T. and Betz, J. (2011) **Queen of the Sun: What are the Bees Telling Us?** Clairview Books.

Steiner, R. (1999) **Bees: nine lectures on the nature of bees**. SteinerBooks, Inc. Transl. Thomas Braatz of Über das Wesen der Bienen, Rudolf Steiner Verlag.

Storch, H. (2014) **At the hive entrance: how to know what happens inside the hive by observation on the outside**. CreateSpace Independent Publishing Platform. Translated from the original Am Flugloch by F. Celis.

Swan, H. (2017) **Where Honeybees Thrive. Stories from the Field**. The Pennsylvania State University Books.

Tautz, J. (2008) **The buzz about bees: biology of a superorganism**. Springer.

Thun, M. (2020) **Biodynamic beekeeping: a sustainable way to keep happy, healthy bees**. Floris Books.

Weiler, M. (2019) **The secrets of bees: an insider's guide to the life of honeybees**. Floris Books.

Film documentaries

Queen Of The Sun
http://www.queenofthesun.com

More Than Honey
*http://www.morethanhoneyfilm.com/?gclid=Cj0KCQjwk8b7BRCaARIsAARRTL7dSxqJ
xHcwedMsd54kgU8WAwCIdSRbRzvn5GUif-d7kaXH2zmGgwwaAsK2EALw_wcB*

HoneyLand
https://honeyland.earth

Beewildered Companions (Mindjazz Pictures)
Streaming 48 h or unlimited: https://vimeo.com/ondemand/beewilderedcompanions.

www.ingramcontent.com/pod-product-compliance
Lightning Source LLC
Chambersburg PA
CBHW051440270326
41932CB00024B/3376